HARCOURT Math

Practice for the TerraNova

Grade 6

Orlando • Boston • Dallas • Chicago • San Diego
www.harcourtschool.com

Copyright © by Harcourt, Inc.

All rights reserved. No part of this publication may be reproduced or transmitted in any form or by any means, electronic or mechanical, including photocopy, recording, or any information storage and retrieval system.

Teachers using HARCOURT MATH may photocopy complete pages in sufficient quantities for classroom use only and not for resale.

HARCOURT and the Harcourt Logo are trademarks of Harcourt, Inc.

TerraNova is a trademark of The McGraw-Hill Companies, Inc. The publisher of TerraNova, CTB/McGraw-Hill, has not endorsed this booklet.

Printed in the United States of America

ISBN 0-15-333653-6

1 2 3 4 5 6 7 8 9 10 082 10 09 08 07 06 05 04 03 02 01

CONTENTS

▶ **Daily Practice**
   Weeks 1–28 4

▶ **Performance Task Practice**
   Test 1 116
   Test 2 126
   Test 3 136

▶ **Answer Pages** 145

▶ **Mathematics Reference Sheet** 157

Name _____

Daily Practice Week 1

1 On January 7, at 7:00 A.M., the temperature in Minneapolis was ⁻4 °F. At 12:00 noon, the temperature rose to ⁺10 °F. Which kind of number expresses these temperatures?

- Ⓐ decimals
- Ⓑ integers
- Ⓒ whole numbers
- Ⓓ fractions

Test Taking Tips

What do you know about whole numbers, integers, fractions, and decimals that will help you eliminate incorrect answers?

2 Bill wants to help his dad put a fence around a garden. He drew a diagram of the garden to help figure out the perimeter. Use the diagram to find the number of feet in the perimeter of the garden.

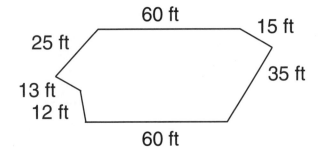

- Ⓕ 200 ft
- Ⓖ 210 ft
- Ⓗ 220 ft
- Ⓘ 230 ft

Test Taking Tips

What is perimeter?

4 Test Prep

Name _____

3 Draw a diagram of a quadrilateral that has 4 right angles and 4 congruent sides.

Name the quadrilateral you drew.

_____

Test Taking Tips

What is a quadrilateral? What is a right angle? What are congruent sides?

---

4 Kirsten led her basketball team in 3-point shots made during the season. She scored a total of 153 points on 3-point shots.

Part A

Write a number sentence that you could use to find the number of 3-point shots that Kirsten made.

_____

Part B

Solve the number sentence that you wrote in Part A. How many 3-point shots did Kirsten make? Explain how you found your answer.

_____

_____

Test Taking Tips

How can knowing the total of points that Kirsten made from 3-point shots help you solve the problem?

Test Prep

Name _____

5 Maria works in a video store. She made a table to record the number of copies of her favorite video that were sold each week for 6 weeks.

Maria's Favorite Videos	
Week	Number Sold
1	15
2	7
3	8
4	4
5	6
6	9

Part A

On the grid on the next page, make a line graph for the data shown in the table. Be sure to

- write a title
- label the graph
- use an appropriate scale
- plot the points
- connect the points

Part B

Explain how you decided what scale to use for the graph.

Part C

Explain why a line graph is an appropriate graph for displaying this data.

Test Taking Tips

How does knowing the range of a set of data help you decide on an interval for the scale?

Name _____

5 Part A

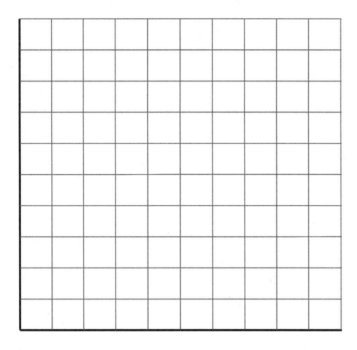

Part B

Explain your scale.

_____

_____

_____

Problem C

Explain why a line graph is appropriate.

_____

_____

_____

_____

Test Taking Tips

How can you check that your answers are accurate?

How can you check that your explanation is clear and complete?

Name _____

1 Mike conducted a survey to find out whether the students in his class want vanilla, chocolate, or strawberry yogurt at the class party.

Which would be the best way for him to display the data for his classmates to see?

- Ⓐ line graph
- Ⓑ circle graph
- Ⓒ frequency table
- Ⓓ bar graph

Test Taking Tips

Which graph shows the comparison of data from one group to another?

---

2 About $\frac{1}{4}$ of the students at West Middle School bring lunch from home. This is represented in the graph below. What percent of students bring their lunch from home?

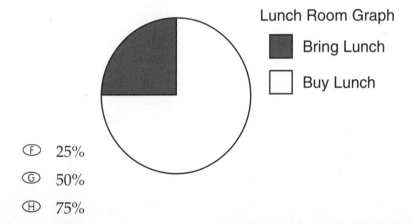

- Ⓕ 25%
- Ⓖ 50%
- Ⓗ 75%
- Ⓘ 100%

Test Taking Tips

If you divide a circle into two equal parts, what percent does each part represent?

If you divide a circle into four equal parts, what percent does each part represent?

Name _____

3 This table shows the circumference and the diameter for some objects Tai measured.

Object	Circumference	Diameter
Spool	9.4 cm	3 cm
Lid	12.6 cm	4 cm
Glass	15.7 cm	5 cm

About how many times longer is the circumference of a circle than its diameter?

_____

Explain how you know.

_____

_____

Test Taking Tips

How does the data in the table help you answer the question?

---

4 In the drawing below name two lines that are parallel and tell why they are parallel.

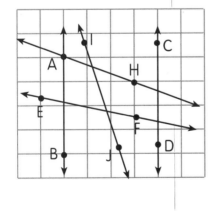

Test Taking Tips

What makes two lines parallel?

_____

_____

_____

Test Prep

Name _____

5 Thomas works in a bakery.

He started this table to show the number of pound of apples he needs to bake medium apple pies.

Medium Apple Pie

Pounds of Apples	3				
Number of Pies	1	2	3	4	5

Part A

On the next page, complete the table for Thomas.

Part B

Thomas needs 4 pounds of apples to make a large apple pie. Make a table to show the number of pounds of apples he needs to make 1, 2, 3, 4, or 5 large apple pies.

Part C

Explain how you completed the tables. How are the two tables the same? How are they different?

Test Taking Tips

How does knowing Thomas needs 3 pounds of apples to make 1 pie help you know how to complete the table?

Name _____

5 Part A

Complete the table.

Medium Apple Pie

Pounds of Apples	3				
Number of Pies	1	2	3	4	5

Part B

Make a table to show how many apples Thomas needs to make a large apple pie.

Part C

Use the data in the table to explain your answers.

_____

_____

_____

_____

_____

Test Taking Tips

How can you check that your answers are accurate?

How can you check that your explanation is clear and complete?

Name _____

1 What is the value of 10^7?

Ⓐ 70
Ⓑ 1,000,000
Ⓒ 10,000,000
Ⓓ 7,000,000,000

Test Taking Tips

What does the exponent tell you about the number of zeroes in the expression written in standard form?

---

2 The table shows the average daily temperature for a 5-day period. How many degrees difference is there in temperature between the highest average daily temperature and the lowest average daily temperature?

Ⓕ 4°
Ⓖ 8°
Ⓗ 9°
Ⓘ 10°

Average Temperature	
Day	Temperature (in °F)
Mon.	60°
Tue.	59°
Wed.	58°
Thu.	51°
Fri.	56°

Test Taking Tips

What is the highest temperature? What is the lowest temperature?

Name _____

Daily Practice Week 3

3 Jonathan weighed his puppy on the first day of the month for 6 months. He made a line graph of the data.

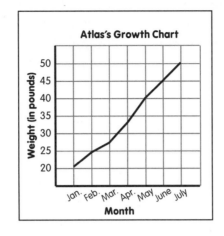

Part A

How much weight did the puppy gain between January 1 and July 1?

_____

Part B

About how much weight did the puppy gain each month? Explain how you found your answer.

_____
_____

Test Taking Tips

What does the graph tell you about the puppy's weight? How can you use this information to solve the problem?

---

4 Mr. Johnson likes to get the best buy when he buys gas. He paid $16.20 for 12 gallons of gas at Pay-n-Save Gas. He paid $20.10 for 15 gallons of gas at Gas Mart. Compare the two prices. Decide which is the better buy and why. Show your work.

_____
_____
_____
_____
_____
_____

Test Taking Tips

What units do you use to measure gas prices?

Test Prep

Name _____

5 Anthony is following directions to make a kite. The directions show a square with vertices labeled *A*, *B*, *C*, and *D*. Use the diagram to help you make a model of part of the kite.

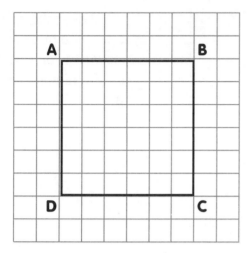

Part A

On the grid on page 15, draw a large square. Label the vertices to match the diagram. Draw line segment *AC*. Describe the new polygons you have formed.

Part B

Now draw a large rectangle. Label the vertices *E*, *F*, *G*, *H*. Draw line segment *EG*. Describe the polygons you have formed.

Part C

Explain how the polygons you drew in the square and the polygons you drew in the rectangles are alike. Explain how they are different.

What is made when you draw a line segment through a square?

Test Prep

Name _____

5 Part A

Draw a large square and line segment in the grid.

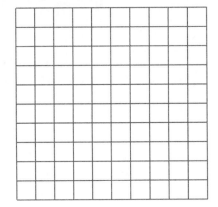

Describe the new polygons you have formed.

_____

Part B

Draw a rectangle and line segment in the grid.

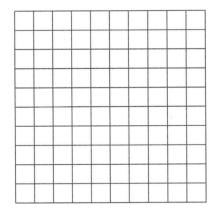

Describe the polygons you have formed.

_____

Part C

Explain the similarities and differences.

_____

_____

_____

Test Prep

Name _____

1 Andrew is playing a math game. He knows one angle of the triangle is 90°. Another angle is 45°. What is the measure of the third angle?

- Ⓐ 45°
- Ⓑ 60°
- Ⓒ 90°
- Ⓓ 180°

Test Taking Tips

What is the sum of the angle measures in a triangle?

---

2 In the table below x represents the number of miles driven by a car and y represents the number of gallons of gas used. What is the missing value for y?

x	1	2	3	4
y	15	30	45	

- Ⓕ 50
- Ⓖ 60
- Ⓗ 70
- Ⓘ 90

Test Taking Tips

What number is x multiplied by to get y?

Name _____

Daily Practice Week 4

3 Anthony recorded his spelling scores for the first 6 weeks of school. The scores were 95, 100, 85, 80, 100, and 75. Find the range of these scores and explain what the range is.

Test Taking Tips

How do you find the range of a set of data?

_____
_____
_____

4 In Juanita's class, 7 of the 28 students are girls. Carlos said $\frac{1}{4}$ of the students were girls. Juanita said only 0.25 of the students were girls. Who is right? Explain your answer.

Test Taking Tips

According to Carlos, how many students are girls?

_____
_____
_____

Test Prep

Name _____

5 Alex wants to cover some boxes with contact paper. One box is a cube with edges of 9 inches. He needs to know the sum of the area of the faces in order to find the total amount of paper he needs. Find the amount of contact paper Alex needs for this box.

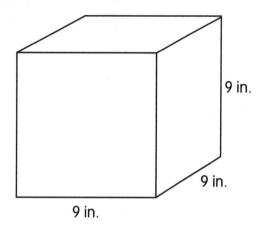

Test Taking Tips

How many faces does the box have?

Part A

Find the area of one face of the cube. Find the sum of the areas of faces. How much contact paper does Alex need?

Part B

Alex has another box that is a cube with sides of 10 inches. How much contact paper will Alex need to cover this box?

Part C

Explain how you found the sum of the areas of the faces on the cubes.

Test Prep

Name _____

5 Part A

How much contact paper does Alex need to cover a box that is a cube with 9-inch edges?

_____

Part B

How much contact paper does Alex need to cover a box that is a cube with 10-inch edges?

_____

Part C

Explain in words.

_____
_____
_____
_____
_____
_____
_____

Test Taking Tips

How can you check that your answers are accurate?

How can you check that your explanation is clear and complete?

Test Prep

Name _____

1 What is the area of a mural that measures 120 inches by 300 inches?

- Ⓐ 360 square inches
- Ⓑ 3,600 square inches
- Ⓒ 36,000 square inches
- Ⓓ 360,000 square inches

Test Taking Tips

How can you use mental math to solve the problem?

---

2 Anna is drawing a pattern to make a birdhouse. The front and the back of the birdhouse will be equilateral triangles. In an equilateral triangle all the angles are congruent.

How many degrees are in each angle of an equilateral triangle?

- Ⓕ 45°
- Ⓖ 60°
- Ⓗ 90°
- Ⓘ 180°

Test Taking Tips

What is the sum of the measures of the angles of a triangle?

Name _____

3 Look at the pattern in the numbers below.

    1, 3, 6, 10, 15

Part A

Write the next three numbers in the pattern.

_____

Part B

Describe the pattern you found.

_____

_____

_____

_____

Test Taking Tips

How is each number in the pattern related to the number before it in the pattern?

4 The chart shows Kevin's scores on his last 6 math quizzes.

What is the mean for his scores?

_____

Explain what this mean tells you.

_____

_____

_____

Test Scores	
Quiz 1	92
Quiz 2	78
Quiz 3	99
Quiz 4	68
Quiz 5	87
Quiz 6	98

Test Taking Tips

How do you find the mean of a set of data?

Name _____

Daily Practice Week 5

5 At the circus, the people in every sixth row got free balloons.

The people in every fifteenth row got free ice cream.

Use the hundreds chart on the next page to help you find the rows where people got free balloons and free ice cream.

Part A

Circle the numbers in the hundreds chart for the rows where people got free balloons.

Mark X's on the numbers for the rows where people got free ice cream.

Part B

List the rows where people got free balloons.

List the rows where people got free ice cream.

In which rows did people get both free balloons and free ice cream?

Part C

Explain in words how you solved the problem.

Test Taking Tips

How do you decide which rows get free balloons?

Test Prep

Name _____

5 Part A

Circle the numbers in the hundreds chart to show the rows where people got free balloons. Mark X's on the chart to show the rows where people got free ice cream.

Test Taking Tips

1	2	3	4	5	6	7	8	9	10
11	12	13	14	15	16	17	18	19	20
21	22	23	24	25	26	27	28	29	30
31	32	33	34	35	36	37	38	39	40
41	42	43	44	45	46	47	48	49	50
51	52	53	54	55	56	57	58	59	60
61	62	63	64	65	66	67	68	69	70
71	72	73	74	75	76	77	78	79	80
81	82	83	84	85	86	87	88	89	90
91	92	93	94	95	96	97	98	99	100

Part B

List the rows where people got free balloons.

_____

List the rows where people got free ice cream.

_____

List the rows where people got both free balloons and free ice cream.

_____

Part C

Explain in words.

_____

_____

_____

Test Prep

Name _____

Daily Practice Week 6

1 Out of 100 tosses of a coin, 47 tosses came up as heads. What percent came up heads?

- Ⓐ 47%
- Ⓑ 52%
- Ⓒ 53%
- Ⓓ 74%

Test Taking Tips

Out of 100 tosses of a coin, what fraction of the tosses came up heads?

2 The distance from Denver to Los Angeles is approximately 1100 miles.

If it takes a plane two hours to fly this distance, how many miles per hour is the plane traveling?

- Ⓕ 50
- Ⓖ 55
- Ⓗ 500
- Ⓘ 550

Test Taking Tips

What does miles per hour mean?

24   Test Prep

Name _____

3 Many things in nature have symmetrical characteristics. The butterfly below is an example.

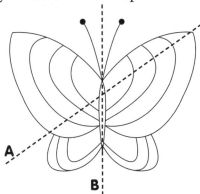

Tell which line is a line of symmetry, and tell how you know the butterfly is symmetrical.

_____

_____

Test Taking Tips

On which line could you fold the butterfly so that it would reflect on itself?

---

4 For a whole school year, Anna recorded the number of hours she studied for all her math exams and the grades she got on each one. Without studying at all, she got a score of 70; studying for 1 hour, she got an 80; studying for 2 hours, she got a 90; and studying for 3 hours, she got a 100.

Part A

Make a table to show the data that Anna recorded.

Part B

Explain how the number of hours Anna studied was related to her test scores.

_____

Test Taking Tips

What do you notice about the test scores as Anna's study time increases?

Test Prep

Name _____

5 The table shows Thomas's social studies test scores for his last seven tests.

Test	1	2	3	4	5	6	7
Score	67	49	75	50	85	80	49

His teacher said he had a choice of using the range, mean, median, or mode of his scores as his average grade for the tests.

This is the work you will do on the next page.

Part A

Find the range mean, median, and mode for the set of data. Which number should he use as his average grade?

Part B

Explain in words how you found the range, mean, median, and mode. Justify the number you chose as Thomas's average grade.

How do you find the range, mean, median, and mode for a set of data?

26   Test Prep

Name _____

5 Part A

Find the range, mean, median, and mode for the set of test scores. Which number should he use?

How can you check that your answers are accurate?

How can you check that your explanation is clear and complete?

Part B

Explain in words.

_____

_____

_____

_____

_____

_____

Test Prep

Name _____

1 Maria knows the *range, mean, median* and *mode* are used to represent the numbers in a set of data.

Which definition below is the definition for the *mean*?

- Ⓐ the difference between the greatest and least numbers in a set of data
- Ⓑ the average of a group of numbers
- Ⓒ the number that occurs the most often in a group of numbers
- Ⓓ the middle number in a group of numbers when arranged in order

Test Taking Tips

Which definition best describes the mean?

---

2 Peninsula Citrus Company is hauling truckloads of citrus to the rest of the country. A driver for one of the trucks told the weigh station he was carrying 480 cases of oranges with 2 dozen oranges per case.

How many oranges is the truck carrying?

- Ⓕ 240 oranges
- Ⓖ 960 oranges
- Ⓗ 5,760 oranges
- Ⓘ 11,520 oranges

Test Taking Tips

How many oranges are in each case?

Name _____

3 Bill receives $5.00 per week for allowance. He is trying to save some money each week for a new CD player. If he saves $1.00 the first week, $1.50 the second week, $2.00 the third week and continues in this way, how many weeks would it take before he needed to save his entire allowance?

Create a table showing this information and how much he saved each consecutive week until he had to save his entire allowance.

Test Taking Tips

How much is his savings increasing from the previous week's savings?

---

4 Diane has a number cube with sides numbered from 1 to 6. She tossed the cube 18 times.

Part A

List the possible outcomes of this experiment.

_____

_____

Part B

Write a fraction for the probability of tossing an even number.

Test Taking Tips

How many of the possible outcomes are even numbers?

Test Prep

Name _____

5 Jose built a storage box that is 2 feet long, 2 feet wide, and 2 feet high. He needs to build a box that has a volume four times greater than the volume of this box.

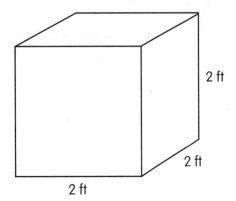

This is the work you will do on the next page.

Part A

Find the volume of the box with 2-foot edges.

Part B

Draw and label boxes with dimensions that have four times the volume of the smaller box.

Part C

Explain in words how changing the length, width, and height of the box changes its volume.

Test Taking Tips

How do you find the volume of a prism?

Name _____

5 Part A

Find the volume of the box with 2-foot edges.

Part B

Draw boxes which have a volume four times the volume of the smaller cube. Do not forget to label the dimensions.

Part C

Explain in words how changing the length, width, and height of the box changes its volume.

_____

_____

_____

Test Taking Tips

How can you check that your answers are accurate?

How can you check that your explanation is clear and complete?

Test Prep

Name _____

1 When Tony buys gasoline, $0.35 of every $1.00 he spends is for taxes. Tony made a table to show how much of every dollar he spends goes for taxes.

Gasoline Prices

Gasoline	$1.00	$2.00	$3.00	$4.00
Tax	$0.35	$0.70	$1.05	x

When Tony spends $4.00 for gasoline, how much of the money goes for taxes?

Ⓐ $1.35
Ⓑ $1.40
Ⓒ $1.70
Ⓓ $2.40

How are the numbers in the table related?

---

2 On the first day of each month, Jody and her grandfather check in the newspaper to find the stock price for Better-Made Toy Company. The prices for the last five months were 105, 86, 70, 90, and 97.

What was the median price for the stock?

Ⓕ 70
Ⓖ 89.6
Ⓗ 90
Ⓘ 105

How do you find the median?

Test Prep

Name _____

Daily Practice Week 8

3 Phil has a job that pays 2 cents on the first day of work. Then for each day after the first, he receives double the preceding day's wages.

Part A

Using exponent form, write the number of cents he will receive on the sixth day.

Part B

Write the value of the number with the exponent. Explain your work.

_____

_____

_____

Test Taking Tips

What operation do you use to find the value of a number with an exponent?

4 A map scale shows that 1 cm on the map equals 15 km in actual distance. Juanita measured 3.7 cm between Center City and Marysville. What is the actual distance between the two cities? Explain your reasoning.

_____

_____

_____

Test Taking Tips

How can you use the scale and Juanita's measure to find the actual distance?

Test Prep 33

Name _____

5 Carlos wants to find lines of symmetry in this figure.

Part A

Trace the figure. Fold it in half in different ways so that the two halves are congruent. Draw a line on the figure for each fold line that represents a line of symmetry.

Part B

Draw another figure that has one or more lines of symmetry. Draw the lines of symmetry for the figure.

Part C

Explain how you know a line drawn on a figure is a line of symmetry.

How can you fold the figure so that the two halves are congruent?

Test Prep

Name _____

5 Part A

Draw the lines of symmetry you found on this small star. How many lines did you find?

_____

Part B

Draw another figure and show the lines of symmetry.

Part C

Explain how to determine the lines of symmetry in a figure.

_____

_____

_____

_____

Test Taking Tips

How can you check that your answers are accurate?

How can you check that your explanation is clear and complete?

Test Prep

Name _____

1 Kayla is studying tessellations in class. She knows a tessellation is a repeating arrangement of shapes that completely covers a plane with no gaps and no overlaps. Which of these shapes would be used to form a tessellation?

Ⓐ   Ⓑ

Ⓒ   Ⓓ

Test Taking Tips

Why can't you use a circle to form a tessellation?

---

2 Catherine is trying to save some money. She kept track of the hours she worked babysitting and the money she earned. The table shows the money she has earned.

Catherine's Babysitting Money				
1	2	3	4	5
$3.00	$6.00	$9.00	$12.00	$15.00

Look at the pattern in the table. How many hours would Catherine need to work in order to earn $30.00?

Ⓕ 4 hours

Ⓖ 8 hours

Ⓗ 9 hours

Ⓘ 10 hours

Test Taking Tips

What does Catherine earn for 1 hour of babysitting?

Test Prep

Name _____

Daily Practice Week 9

3 The doctor told Carla's mother that the probability of having twins is $\frac{1}{50}$ and the probability of having a boy is $\frac{1}{2}$. Based on these statistics, which is more likely: having twins or having a boy?

_____

Explain your answer.

_____
_____
_____
_____

 **Test Taking Tips**

What is the definition of probability?

4 Kelley is frustrated with math homework. The teacher marked the problem as wrong, but Kelley keeps coming up with the same answer. Study the problem and Kelley's answer. Write the problem with the correct answer on your answer sheet. Explain what Kelley was doing wrong.

$4 + 3 \times 5 = x$

Kelley keeps getting the answer 35.

_____
_____
_____

 **Test Taking Tips**

What do you get if you add and then multiply?

What do you get if you multiply and then add?

Test Prep

Name _____

5 The sixth graders are having a pizza sale to make money for their outdoor-lab school. They have decided to make square pizzas. A pizza with 12-inch sides sells for $8.00. They need to decide how much to charge for 6-inch and 18-inch pizzas.

How does finding the area of each pizza help you decide what to charge for the 6-inch and the 18-inch pizza?

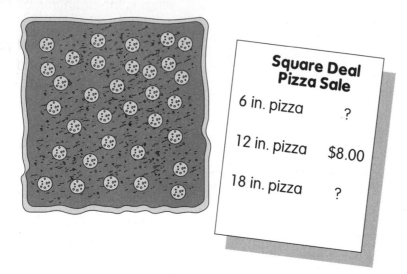

Part A

On the next page, draw diagrams of the three square pizzas. Label the lengths of sides of each pizza. Find the area of each pizza. Explain how the areas are related.

Part B

Use the relationship between the areas of the pizzas to determine a price for the 6-inch and th 18-inch pizza.

Part C

Explain how you decided to charge for the pizzas.

Test Prep

Name _____

5 Part A

Draw diagrams of the three square pizzas. Be sure to label the sides and find the area of each.

Part B

Use the price of the 12-inch pizza and the relationship between the areas of the pizzas to decide on a price for the 6-inch pizza and the 18-inch pizza.

Part C

Explain in words.

_____

_____

_____

Test Prep

Test Taking Tips

How can you check that your answers are accurate?

How can you check that your explanation is clear and complete?

Name _____

1 There is a sale at Sound Wherehouse. CD's are 3 for $43.50. Carol wants to know the price of one CD so she can compare prices.

Which of these expressions can Carol use to find the cost of one CD?

- Ⓐ $43.50 × 3
- Ⓑ $43.50 ÷ 3
- Ⓒ 3 ÷ $43.50
- Ⓓ $43.50 + 3

Test Taking Tips

Would the cost of one CD be more or less than the cost of 3 CDs?

---

2 Jason knows that perpendicular lines intersect. How many degrees are in the angles formed by perpendicular lines?

- Ⓕ 45°
- Ⓖ 60°
- Ⓗ 90°
- Ⓘ 180°

Test Taking Tips

What geometric figure has angles made by perpendicular lines?

Name _____

Daily Practice Week 10

3 Christopher needs to time the runners who are running a 10-kilometer race in San Francisco. He needs to record accurate times for this race because he knows several runners often come in very close to the same times. What would be a good instrument for him to use? Explain why.

Test Taking Tips

What instrument gives accurate time readings for seconds and parts of a second?

4 The weather forecaster said the probability of rain today is $\frac{1}{5}$. What is the probability that it will NOT rain today?

_____
_____
_____
_____
_____
_____
_____

Test Taking Tips

What does it mean to say the probability of rain is $\frac{1}{5}$?

Test Prep

Name _____

5 As a homework assignment, Wendy is writing "real-life" problems for these expressions.

$1{,}170 \div 45$ $\dfrac{7}{8} \times 2$

$4\dfrac{3}{4} - \dfrac{7}{8}$ $10.75 + 15.49$

Test Taking Tips

When do you use each operation in problems?

For the first expression Wendy wrote this problem.

There are 1,170 marbles in one jar.

How many marbles are in 45 jars?

Part A

Did Wendy write a problem appropriate for the expression? Explain on the next page.

Part B

Help Wendy with her homework. Write a problem for each expression.

Name _____

5 Part A

Explain in words.

_____

_____

_____

_____

Part B

Write a "real-life" problem for each expression.

$1{,}170 \div 45$

$\frac{7}{8} \times 2$

$4\frac{3}{4} - \frac{7}{8}$

$10.75 + 15.49$

Test Taking Tips

How can you check that your answers are accurate?

How can you check that your explanation is clear and complete?

Name _____

Daily Practice Week 11

1 The Pizza Pantry is having a contest. Jerrod can get a bargain pizza if he finds the row where all four numbers are equivalent. Which row should Jerrod pick?

Ⓐ $\frac{1}{4}$ .25 $\frac{2}{5}$ 25%

Ⓑ $\frac{1}{5}$ .20 $\frac{2}{10}$ 20%

Ⓒ $\frac{2}{4}$ .50 $\frac{6}{10}$ 50%

Ⓓ $\frac{8}{10}$ .40 $\frac{4}{5}$ 40%

Test Taking Tips

How could writing all the numbers in a row as fractions help you solve the problem?

2 On his last five math tests, Chin's scores were 100, 97, 99, 90, and 100. What is his mean score?

Ⓕ 97

Ⓖ 97.2

Ⓗ 99

Ⓘ 100

Test Taking Tips

How do you find the mean?

Grade 6 • Harcourt

44   Test Prep

Name _____

3 Philip wants to find the area of a rectangular parcel of land. Because of the thick foliage, he can only measure the length of one side (*CD*). He knows that the length of *AD* is twice the length of *CD*.

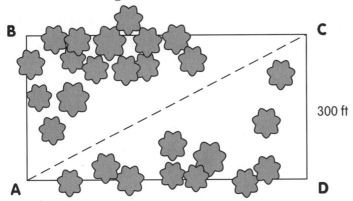

Find the area of the parcel of land. Show your work or explain in words how you found the answer.

_____

_____

Test Taking Tips

How long is the line segment *AD*?

4 Lynn plans to run in a 10-mile race to raise money for charity. Sponsors agree to donate a certain amount for every mile that she runs. One sponsor agrees to donate an amount equal to two times the square of the number of miles that she runs. The table shows the donations the sponsor will make if Lynn runs 1 mile or 2 miles.

Test Taking Tips

What does "two times the square of a number" mean?

Sponsor's Donations							
Miles Run	1	2	3	4	5	6	7
Donation	$2	$8					

Complete the table to show the amount the sponsor will donate for the other miles shown.

Test Prep

Name _____

Daily Practice Week 11

5 Jaime's little sister is selling lemonade. She has cups in three sizes. The small cups hold 6 ounces, the medium cups hold 12 ounces, and the large cups hold 15 ounces. Jaime is helping her sister decide how much to charge for the cups of lemonade. They decide to charge $0.40 for a medium cup.

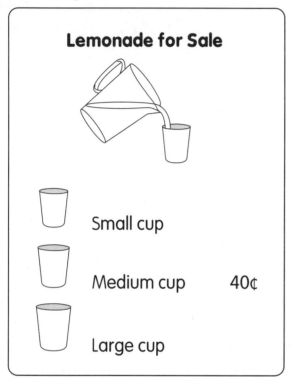

Test Taking Tips

How does knowing the capacity of each cup help you decide on a price?

Part A

On the next page, draw diagrams of the three cups. Write the number of ounces each cup holds on the cup. Explain how the capacities are related.

Part B

Use the relationship between the capacities of the cups to help the girls determine a price for the small cup and the large cup of lemonade.

Part C

Explain how you decided what to charge for each cup.

Test Prep

Name _____

5 Part A

Draw diagrams of the three cups. Be sure to label the cups.

Test Taking Tips

How can you check that your answers are accurate?

How can you check that your explanation is clear and complete?

Part B

Use the price of the medium cup of lemonade and the relationship between the capacities of the cups to decide on a price for the small cup and the large cup of lemonade.

Part C

Explain your reasoning.

_____

_____

_____

_____

Test Prep

Name _____

Daily Practice Week 12

1 The mean monthly temperatures for the last six months in Phoenix, Arizona, are listed below.

Temperatures	
January	78°F
February	76°F
March	83°F
April	90°F
May	92°F
June	97°F

What is the mean temperature for the first 6 months of the year in Phoenix, Arizona?

Ⓐ 76°
Ⓑ 83°
Ⓒ 85°
Ⓓ 86°

Test Taking Tips

How do you find the mean?

---

2 Kevin loves granola bars. If Kevin wants to triple the recipe below, how many cups of flour will he need?

Ⓕ $\frac{1}{6}$ c

Ⓖ $\frac{1}{8}$ c

Ⓗ 1 c

Ⓘ $\frac{3}{2}$ c

1 cup granola
1 cup quick-cooking rolled oats
1 cup chopped nuts
$\frac{1}{2}$ cup flour
$\frac{1}{2}$ cup raisins
1 beaten egg
$\frac{1}{3}$ cup honey
$\frac{1}{3}$ cup cooking oil
$\frac{1}{4}$ cup packed brown sugar
$\frac{1}{2}$ tsp ground cinnamon

Test Taking Tips

What does it mean to triple a recipe?

---

48 Test Prep

Name _____

3 The Smiths want to find out how much wallpaper they will need to cover the four walls in their dining room. The dining room floor is 9 feet by 12 feet. It is 10 feet from the ceiling to the floor in the dining room.

Part A

What is the area of each wall?

Part B

What is the total area of the four walls?

Test Taking Tips

How would you find the area of a wall?

---

4 Draw a 90° clockwise rotation of the figure about the point of rotation.

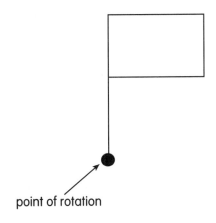

point of rotation

Test Taking Tips

How can tracing the figure help you show the rotation?

Test Prep

Name _____

5 Beth has a 3-inch by 4-inch photo that she wants to enlarge and place in a frame that has a 12-inch by 12-inch opening. The original photo and the enlargement will be similar rectangles.

Test Taking Tips

What do you know about the ratios of the lengths of matching sides in similar figures?

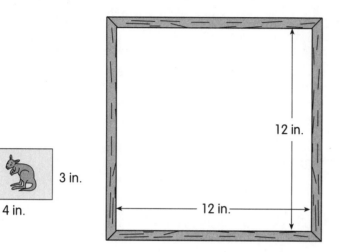

Here is the work you will do on the following page.

Part A

What are the dimensions of the largest enlargement Beth can fit in the frame? Draw a diagram to show the photo in the frame. Write the dimensions on the photo.

Part B

Suppose Beth wants to leave a space that is at least 2 inches on all sides between the photo and the frame. What are the dimensions of the largest enlargement Beth can use now? Draw a diagram to show the photo in the frame. Write the dimensions on the photo.

Part C

Explain how you found the dimensions of the photos.

Test Prep

Name _____

5 Part A

What are the dimensions of the largest enlargement Beth can fit in the frame? Don't forget to draw and label a diagram of the photo and frame.

Part B

What are the dimensions of the largest enlargement Beth can fit in the frame if she leaves at least 2 inches on all sides between the photo and the frame? Don't forget to draw and label a diagram of the photo and frame.

Part C

Explain in words.

_____

_____

_____

_____

Test Taking Tips

How can you check that your answers are accurate?

How can you check that your explanation is clear and complete?

Test Prep

Name _____

Daily Practice Week 13

1 Jean and Scot are building a tree house with plywood sheets and pine boards. Their plan calls for three 4 ft by 8 ft plywood sheets for every eight 2 in. x 4 in. x 6 ft pine boards. They determine that they need 9 sheets of plywood.

How many pine boards do they need?

- Ⓐ 8
- Ⓑ 14
- Ⓒ 24
- Ⓓ 28

Test Taking Tips

What is the relationship between the number of sheets of plywood and the number of pine boards?

2 For the last week the weather forecaster has been saying that the probability of rain is $\frac{1}{2}$. It has not rained at all.

He says the probability that it will rain tomorrow is $\frac{3}{4}$. What is the probability that it will NOT rain tomorrow?

- Ⓕ 0
- Ⓖ $\frac{1}{4}$
- Ⓗ $\frac{1}{2}$
- Ⓘ $\frac{3}{4}$

Test Taking Tips

What information is given in the problem that you don't need to solve the problem?

52 Test Prep

Name _____

③ Christine needs new shoes. There is a special sale on at Crosby's Shoe Store. The sign reads as follows:

Labor Day Shoe Sale
All shoes $\frac{1}{2}$ off!

Christine found 3 pairs she would like. She is hoping that at $\frac{1}{2}$ off she will have enough money to buy all 3 pairs. The original price for the first pair is $49.50. The second pair's original price is $38.00 and the third pair was marked $35.00. Christine has $60.00.

Can she buy all 3 pairs of shoes? Explain.

_____

_____

Test Taking Tips

How can you find the price of a pair of shoes if they are $\frac{1}{2}$ off the marked price?

④ A camel weighs between 450 kg and 725 kg. He can carry a load of up to 445,000 grams. How many kilograms can a camel carry? Show your work and explain.

_____

_____

_____

_____

_____

_____

Test Taking Tips

What data do you need to solve the problem?

Test Prep

Name _____

5 Tessellations are designs that are found in nature, in art, and in manufactured objects. The design is a repeating arrangement of shapes that completely covers a plane with no gaps and no overlaps.

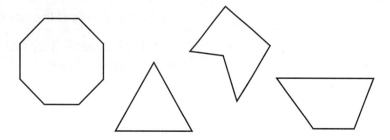

What is a tessellation?

Part A

Trace each shape above several times and cut out your tracings. On the next page, circle the two shapes that can be used to form tessellations.

Part B

Draw a design that makes a tessellation. Use one of the shapes above or a different shape. Make at least three rows in your design.

Part C

Explain how you know your design forms a tessellation.

Name _____

5

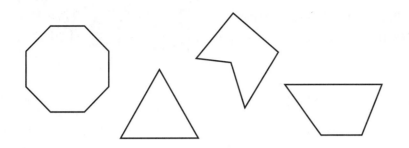

Part A

Which two shapes can be used to form tessellations?

_____

_____

Part B

Draw a design that makes a tessellation.

Part C

Explain in words.

_____

_____

_____

_____

Test Taking Tips

How can you check that your answers are accurate?

How can you check that your explanation is clear and complete?

Name _____

1 Flip the triangle across the dashed line. What is the ordered pair for the NEW location of point *A*?

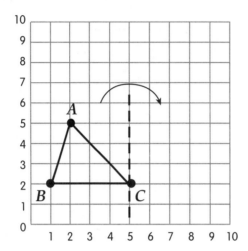

- Ⓐ (2, 5)
- Ⓑ (5, 2)
- Ⓒ (8, 5)
- Ⓓ (5, 8)

What will happen to point C when the triangle is flipped?

2 Terry's great grandmother is 81 years older than Terry. Terry is half the age of his brother Todd. Todd is 14 years old. How many years old is Terry's great grandmother?

- Ⓕ 88
- Ⓖ 89
- Ⓗ 95
- Ⓘ 109

How old is Terry?

Name _____

3

David's Report Card Grades	
Math	97%
English	95%
Science	90%
Soc. Studies	98%
Phys. Ed	75%

David's uncle saw his report card and said he was impressed with David's grades and his overall average grade. What is David's overall average grade?

_____

Explain how you would interpret his individual grades and his overall average grade.

_____

_____

_____

How do you find mean or average?

Test Taking Tips

4 Contest! Guess how many apples are in the barrel, and win them. Lisa made a guess. What would be a good estimate?

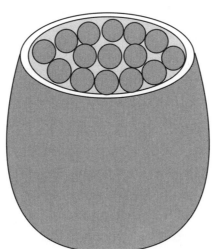

_____

Tell why.

_____

_____

Test Taking Tips

How many layers of apples could there be underneath?

Test Prep

Name _____

Daily Practice Week 14

5 Edward and his dad are putting together a new bicycle for Edward's little brother. There is a scale drawing in the instructions. It says the scale is $\frac{1}{2}$ inch = 1 foot.

Test Taking Tips

Scale: $\frac{1}{2}$ inch = 1 foot

If the scale is $\frac{1}{2}$ inch = 1 foot, how do you find the actual length of something that measures 1 inch in a diagram?

A scale of $\frac{1}{2}$ inch = 1 foot means that every $\frac{1}{2}$ inch in drawing of the bicycle represents 1 foot of actual size.

On the next page, you will find the actual size of the bicycle they are building. The steps are listed below.

Part A

Use an inch ruler to measure the height and the length of the bike in the diagram.

Part B

Use the height and length of the bicycle in the diagram and the scale to find the actual height and length of the new bicycle.

Part C

Explain how you used the diagram and the scale to find the actual size of the bicycle. Explain your reasoning.

58 Test Prep

Name _____

5 Part A

Measure the height and the length of the bicycle in the diagram.

       height in drawing: _____

       length in drawing: _____

Part B

Find the actual height and length of the bicycle.

       actual height: _____

       actual length: _____

Test Taking Tips

How can you check that your answers are accurate?

How can you check that your explanation is clear and complete?

Part C

_____

_____

_____

_____

_____

_____

Test Prep

Name _____

Daily Practice Week 15

1 Shelley's family is building a new deck on the back of their house. A carpenter estimated that the wood for a 12 by 12 ft deck would cost $2,400. About how much would the wood for a 24 ft by 24 ft deck cost?

Ⓐ about $3,500

Ⓑ about $4,800

Ⓒ about $7,200

Ⓓ about $9,600

Test Taking Tips

How many times larger than a 12 ft x 12 ft deck is a 24 ft x 24 ft deck?

2

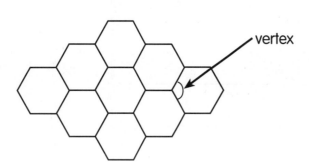

The design above is an example of a tessellation.

How many degrees are there in the sum of the measures of the angles around each vertex inside the design?

Ⓕ 120° Ⓗ 720°

Ⓖ 360° Ⓘ 900°

Test Taking Tips

How many degrees are in a circle?

60

Test Prep

Name _____

3 Look at the pattern of numbers below.

3, 5, 9, 17, 33

Part A

Write the next three numbers in the pattern.

_____

Part B

Write a rule or expression that could be used to find the numbers in the pattern.

_____

_____

_____

Test Taking Tips

What combination of multiplication and subtraction could you use to go from one number in the pattern to the next?

4 Earthquakes in Northern California are common. There may be as many as ten a week. However, most are never felt because their magnitudes are so low. The probability of NOT feeling an earthquake is $\frac{99}{100}$. Does that mean that if 99 earthquakes occur and are not felt that the next earthquake will be felt? Explain.

_____

_____

_____

_____

Test Taking Tips

What does a probability of $\frac{99}{100}$ mean?

Test Prep

Name _____

5 Joseph would like to be a cartoon animator when he grows up. However, he just found out that one animated movie may have up to 345,000 separate drawings. A $3\frac{1}{2}$-second scene takes about 100 drawings.

Joseph started this table to show about how many drawings would be made for every $3\frac{1}{2}$ seconds of animation.

Length of Film (in seconds)	Number of Drawings
$3\frac{1}{2}$	100
7	200
$10\frac{1}{2}$	300
	400
	500
	600
	700
	800
	900
	1,000

Part A

Copy and complete the table. How long is an animated film with 1,000 drawings?

Part B

If it takes 2 hours to complete a drawing, about how many hours will it take to make the drawings for an animated film that is 1 minute and 45 seconds (105 seconds) long?

Part C

Explain how you determined your answers.

Test Taking Tips

How many drawing are drawn for $3\frac{1}{2}$ seconds? For 7 seconds?

Name _____

5 Part A

Copy and complete the table. How long is an animated film with 1,000 drawings?

Test Taking Tips

How can you check that your answers are accurate?

How can you check that your explanation is clear and complete?

Part B

If it takes 2 hours to complete a drawing, about how many hours will it take to make the drawings for an animated film that is 1 minute and 45 seconds (105 seconds) long?

_____

Part C

Explain how you determined your answers.

_____

_____

_____

_____

Test Prep

63

Name _____

Daily Practice Week 16

1 The distance from Earth to the moon is approximately 400,000 km.

Which expression below is equivalent to 400,000?

Ⓐ 4×10^4

Ⓑ $4 \div 10^4$

Ⓒ 4×10^5

Ⓓ $4 \div 10^5$

Test Taking Tips

What is the value of 10^2? of 10^3?

---

2 The Johnsons are building a new home. Mr. Johnson had blueprints drawn up that are scale drawings of the home. The scale on the blueprint is 10 cm = 2 m. Robert measured his room on the blueprint. It was 30 cm long. How many meters long is the actual length of the room?

Ⓕ 6 m

Ⓖ 22 m

Ⓗ 50 m

Ⓘ 150 m

Test Taking Tips

What does a scale of 10 cm = 2 m mean?

Test Prep

Name _____

3 On grid paper, draw the triangle below. Then draw the triangle as it would appear after a slide of 4 units to the right and 3 units up. Label the new triangle *EFG*. Write the ordered pairs for *E*, *F*, and *G*.

What is a slide?

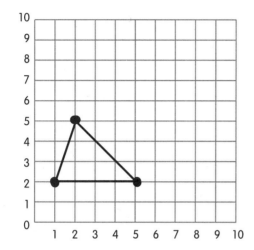

4 Tom wants to write an equation to find out how much of his allowance he spent. His allowance was $8.00. He isn't sure how he spent the money or how much he spent, but he knows he has $0.79 left.

Part A

Write an equation to help Tom find out how much he spent. Let *n* = the amount he spent.

What is an equation?

Part B

Solve the equation to find out how much he spent.

Test Prep

65

Name _____

5 Joe has designed this spinner for a new math board game he and his classmates are making. Before he decides on the rules for moving along the path on the game board, he wants to know the probability of spinning certain numbers.

Why is knowing probability important when you want to design a fair game?

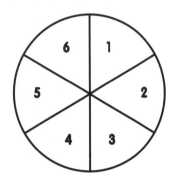

Part A

Use Joe's spinner. Write each probability as a fraction.

a. P (6)

b. P (3 or 4)

c. P (a multiple of 2)

d. P (odd number)

e. P (number less than 7)

f. P (number greater than 7)

Part B

Design a new 6-section spinner. Write numbers in the sections of your spinner so that all these probabilities are true.

a. P (even number) = $\frac{3}{6} = \frac{1}{2}$

b. P (number less than 4) = $\frac{0}{6} = 0$

c. P (7 or 8) = $\frac{2}{6} = \frac{1}{3}$

d. P (number greater than 12) = $\frac{0}{6} = 0$

Part C

Write three more probabilities that are true for the spinner you designed.

Test Prep

Name _____

5 Part A

Write each probability as a fraction.

a. _____   d. _____

b. _____   e. _____

c. _____   f. _____

Part B

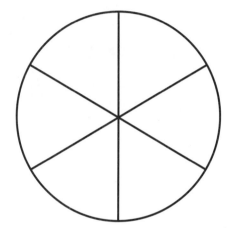

Design the new spinner. Be sure that all four probabilities are true for your spinner.

Part C

Write three more probabilities that are true for the spinner you designed.

_____

_____

_____

_____

Test Taking Tips

How can you check that your answers are accurate?

How can you check that your explanation is clear and complete?

Test Prep

Name _____

1 Phillip is playing a new game with a number cube labeled 1, 2, 3, 4, 5, and 6. Phillip rolled a 5 three times in a row.

What is the probability that he will roll a 5 on his next turn?

Ⓐ 0

Ⓑ $\frac{1}{6}$

Ⓒ $\frac{1}{5}$

Ⓓ $\frac{1}{3}$

Test Taking Tips

What is the probability of rolling a 5 with a cube labeled 1, 2, 3, 4, 5, and 6?

---

2 Douglas County, Colorado, is one of the fastest growing counties in the country. In the last 3 years they have built 9 new elementary schools.

If the county's growth continues at this rate, how many elementary schools will they need to build in the next 10 years?

Ⓕ 27 schools

Ⓖ 30 schools

Ⓗ 60 schools

Ⓘ 90 schools

Test Taking Tips

An average of how many new schools are built each year?

Name _____

3 Peter lives about 45 minutes from the airport. His flight leaves at 2:35 P.M. He knows he must be at the airport 1 hour before the flight for check-in.

What time should he leave for the airport?

_____

Explain your reasoning.

_____

_____

_____

---

4 Rob works as a waiter in a restaurant. He earns an hourly salary plus tips. On Friday night he earned a total of $94 after working an 8-hour shift. His tips for the shift were $70.

Part A

Write an equation that you could use to determine Rob's hourly salary.

_____

Part B

Solve the equation that you wrote in Part A. Show your work or explain how you found your answer in words.

_____

_____

Test Taking Tips

How does working backwards help you solve this problem?

Test Taking Tips

What information does the problem give you? What do you need to figure out to solve the problem?

Name _____

5 Jamaal bakes chocolate chip cookies and sells them to a grocery store. He sells each package of cookies to the store for $4.50. Each package costs Jamaal $2.50 to make and package. The difference between what he sells the cookies for and what they cost him to make and package is his profit. Here is a chart that he started.

Jamaal's Cookie Sales

Month	Packages Sold	Sales	Costs	Profits
January	3	$13.50	$7.50	$6
February	6			
March	5			
April	12			
May	8			
June	15			

Test Taking Tips

What do you need to figure out to solve the problem?

Part A

Complete the table showing Jamaal's cookies sales over a 6-month period.

Part B

Jamaal wants to make a profit of $36 each month. Write and solve an equation to show how many packages of cookies he will have to sell to make a profit of $36.
Let c = the number of packages of cookies he must sell.

70 Test Prep

Name _____

5 Part A

Complete the table showing Jamaal's cookie sales over a 6-month period.

Jamaal's Cookie Sales				
Month	Packages Sold	Sales	Costs	Profits
January	3	$13.50	$7.50	$6
February	6			
March	5			
April	12			
May	8			
June	15			

Part B

Jamaal wants to make a profit of $36 each month. Write and solve an equation to show how many packages of cookies he will have to sell to make a profit of $36.
Let c = the number of packages of cookies he must sell.

Test Taking Tips

How can you check that your answers are accurate?

How can you check that your explanation is clear and complete?

Test Prep

Name _____

Daily Practice Week 18

1 Selene recorded how much time she spent on homework over a 30-day period. She studied 72 hours during the period.

Which equation below could be used to determine the average number of hours she spent on homework each day during the 30-day period?

Ⓐ $x = 30 \div 72$

Ⓑ $30 + x = 72$

Ⓒ $72x = 30$

Ⓓ $x = 72 \div 30$

Test Taking Tips

What is the problem asking you to find out?

Restate the problem in your own words to help you solve it.

---

2 Mara has 6 coins in her coin purse. She knows 3 are pennies, 2 are dimes and 1 is a nickel.

If she reaches in and grabs the first coin she touches, what is the probability, written as a fraction, that she will choose a penny?

Ⓕ $\frac{1}{6}$

Ⓖ $\frac{2}{6}$

Ⓗ $\frac{3}{12}$

Ⓘ $\frac{1}{2}$

Test Taking Tips

What is probability?

Test Prep

Name _____

3 The bell on the town clock rings every hour on the hour. The school bell next door rings every 40 minutes between 8:20 A.M. and 3:00 P.M. How many times will the 2 bells be ringing at the same time each day?

_____

Test Taking Tips

How will making a table help you solve the problem?

Explain how you got your answer.

_____

_____

_____

4 The Sears Tower in Chicago is one of the world's tallest skyscrapers. It is approximately 485 yards tall. In the United States, however, we often speak of heights of buildings in feet.

How many feet tall is the Sears Tower?

_____

Explain your work.

_____

_____

_____

_____

Test Taking Tips

How are feet related to yards?

Test Prep

Name _____

Daily Practice Week 18

Test Taking Tips

⑤

Some Geometric Terms		
ray	line segment	intersecting
angle	parallel	acute
line	perpendicular	right
plane	point	obtuse

People often use geometric terms such as those above to describe figures on a map. Sara drew this map of her neighborhood.

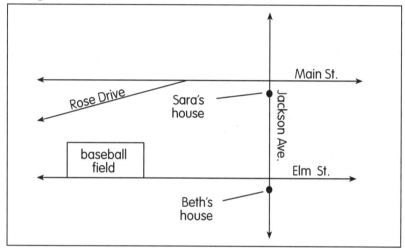

Part A

Use a geometric term to name the figure suggested by each location listed below. You may be able to use more than one term for some examples.

    a. Sara's house
    b. straight path from Beth's house north
    c. path from Sara's house to Beth's house
    d. Main Street intersected by Jackson Avenue
    e. Main Street and Elm Street
    f. baseball field

Part B

Draw a simple map of your neighborhood. Be sure to include figures that can be named as angles and lines or line segments.

Part C

Use at least 6 geometric terms to describe geometric figures on your map.

What object in your classroom reminds you of each of these geometric terms?

    angle
    line segment
    ray

Test Prep

Name _____

5 Part A

List a geometric term suggested by each location.

a. _____

b. _____

c. _____

d. _____

e. _____

f. _____

Part B

Draw a simple map of your neighborhood. Be sure to include figures that can be named as polygons, angles, and line segments.

Part C

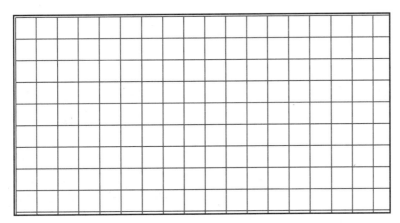

Identify six geometric figures on your map.

_____

_____

_____

_____

Test Taking Tips

How can you check that your answers are accurate?

How can you check that your explanation is clear and complete?

Test Prep

Name _____

1 Triangles *ABC* and *DEF* are right triangles with dimensions as shown below.

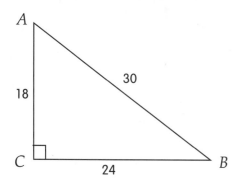
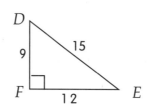

What word best describes the relationship between these triangles?

Ⓐ congruent
Ⓑ symmetrical
Ⓒ similar
Ⓓ parallel

Test Taking Tips

What do the triangles have in common? How are they different?

---

2 Dana sells subscriptions to a newspaper. For every new 6-month subscription she gets $3. For every yearly subscription she gets $6. After one month she sold 10 yearly subscriptions and earned a total of $114.

How many 6-month subscriptions did she sell?

Ⓕ 18 subscriptions
Ⓖ 38 subscriptions
Ⓗ 54 subscriptions
Ⓘ 84 subscriptions

Test Taking Tips

How much did Dana earn selling yearly subscriptions?

Name _____

3 The movie Josh rented was 124 minutes long. He started it at 8:24 P.M.

What time will the movie be over?

_____

Explain your work.

_____

_____

_____

_____

Test Taking Tips

How are minutes and hours related?

---

4 Mary is trying to find the greatest common factor of 20 and 24.

First she listed all the factors of 20: 1, 2, 4, 5, 10, 20.
She then listed all the factors of 24: 1, 2, 3, 4, 6, 8, 12, 24.

What is the greatest common factor of 20 and 24?

_____

Explain what Mary might be doing if she needs to find a greatest common factor.

_____

_____

_____

_____

Test Taking Tips

What are the common factors of 20 and 24?

Test Prep

Name _____

5 Karen made this table to record her scores on math tests during the year.

Karen's Test Scores							
Test	1	2	3	4	5	6	7
Scores	88	90	74	94	100	80	90

She wants to analyze her scores to see how she is doing.

Part A

Make a graph for the data shown in the table. Remember to label the graph and write a title.

Part B

Find the mean, median, and mode of Karen's test scores.

Part C

Explain your choice for the type of graph you made to display Karen's scores. Write a sentence to describe the data.

Test Taking Tips

When is it most appropriate to use a bar graph? a circle graph? a line graph?

Name _____

5 Part A

Decide what kind of graph will best display the data. Make the graph on the grid. Be sure to label the graph and write a title.

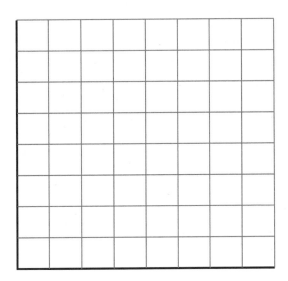

Part B

Find the mean, median, and mode of Karen's scores.

Mean: _____

Median: _____

Mode: _____

Part C

Explain your choice for the type of graph you made to display Karen's scores. Write a sentence to describe the data.

_____

_____

_____

Test Taking Tips

How can you check that your answers are accurate?

How can you check that your explanation is clear and complete?

Name _____

1

The St. Louis Arch is shown above. If the scale for the drawing is 1 in. = 450 ft, and the drawing is $1\frac{1}{2}$ in. high, what is the approximate height of the arch?

Ⓐ between 400 ft and 500 ft
Ⓑ between 500 ft and 600 ft
Ⓒ between 600 ft and 700 ft
Ⓓ between 700 ft and 800 ft

What does it mean to say that the scale is 1 in. = 450 ft?

---

2 A popcorn box is 6 centimeters wide, 12 centimeters long, and 20 centimeters high. It holds 4 cups of popcorn. How many cups of popcorn would a box hold that is 6 centimeters wide, 24 centimeters long, and 20 centimeters high?

Ⓕ 4 c
Ⓖ 8 c
Ⓗ 12 c
Ⓘ 16 c

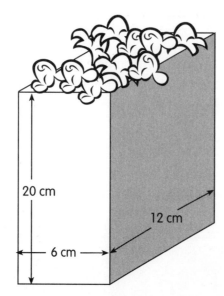

What would happen to the volume of the box if you doubled the size?

Test Prep

Name _____

3 The school newspaper printed the percent of votes that each candidate for class president received. Annette Dawson received 40%, Alan Newman received 25%, Sang Lee received 20%, and Jason Green received 15% of the 150 votes cast. Rene conducted a poll of 20 students in the marching band on how they voted. She found that 10 students voted for Dawson, 5 for Newman, 0 voted for Lee, and 5 voted for Green.

Part A

Find the percent of votes that each candidate received in Rene's poll. How did the results of Rene's poll compare with the actual voting results?

_____

Part B

Explain why you think the results of the poll may have been different.

_____

Test Taking Tips

How do you find the percent one number is of another?

4 Teresa is helping her little sister estimate how many beans are in a 10-pound bag. First they filled a measuring cup with beans and counted 143 beans in one cup.

Using this information, how could you make an estimate of how many beans are in the bag without counting them all one by one? Give details of your strategy and plan.

Test Taking Tips

How could you find out how many cups of beans are in 10 pounds?

_____

_____

_____

Test Prep

Name _____

5 Juan conducted a survey of 20 students in his class about their favorite subjects in school. His survey revealed the following information:

    10 students preferred social studies
    5 students preferred art
    3 students preferred science
    1 student preferred English
    1 student preferred math

Part A

Draw a circle graph to illustrate Juan's survey results. Be sure to label the sections of your graph and write a title.

Part B

Draw a bar graph to illustrate Juan's survey results. Be sure to label your graph and write a title.

Part C

Which graph do you think is easier to understand? Explain your choice.

Test Taking Tips

What type of graph can be used to show how the parts relate to the whole? How do you determine the number of degrees in each section of a circle graph?

Test Prep

Name _____

5 Part A

Draw a circle graph. Remember labels and a title.

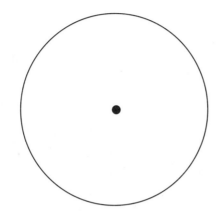

Test Taking Tips

How can you check that your answers are accurate?

How can you check that your explanation is clear and complete?

Part B

Draw a bar graph. Remember labels and a title.

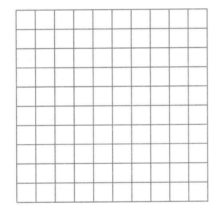

Part C

Which graph do you think is easier to understand? Explain your choice.

_____

_____

Test Prep

Name _____

Daily Practice Week 21

1 Wanda is starting a new exercise program. She was told to start slow and build up gradually.

If she exercises 15 minutes on Monday and increases her time by 10 minutes each day, how long will she be exercising by Friday?

Ⓐ 45 minutes
Ⓑ 55 minutes
Ⓒ 65 minutes
Ⓓ 75 minutes

Test Taking Tips

How long will she exercise on Tuesday?
How long on Wednesday?
What pattern do you see?

---

2 Janet is mixing some juices to make a large bowl of punch. She mixed 3 liters of grape juice, 2.5 liters of apple juice, and 1200 milliliters of cranberry juice.

How many milliliters of juice did she make in all?

Ⓕ 1205.5 mL
Ⓖ 1750 mL
Ⓗ 6205 mL
Ⓘ 6700 mL

Test Taking Tips

How many milliliters are in one liter?

Name _____

3 Every morning, Jane jogs around a small park near her home. The park is bordered by a walkway in the shape of a regular octagon. Jane wants to figure out how many yards she runs each morning. She knows that the length along one side of the park is 175 yards. What is the distance in yards around the whole park?

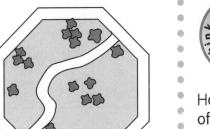

_____

Show your work or explain your answer in words.

_____

_____

_____

Test Taking Tips

How can you use the length of one side of the park to help you find the solution to the problem?

---

4 Rebecca received a $10,000 loan from a bank to start her own business. The interest rate on the loan is 5% per year, simple interest. She plans to begin paying off the loan after 2 years.

Part A

Write an equation that could be used to determine the total amount that Rebecca will owe to the bank after 2 years. (Remember, she will owe the amount of the loan plus the interest on the loan.)

_____

Part B

Solve the equation that you wrote in Part A. Show your work or explain how you found your answer.

_____

Test Taking Tips

What information does the problem give you? What do you need to find out to solve the problem?

Test Prep

Name _____

⑤

Math Test-Score Data				
Tests	Range	Median	Mode	Mean
37, 65, 42, 55, 65	28	55	65	53
58, 42, 75, 85, 45	43	58	–	61
38, 27, 30, 55, 45	28			
45, 35, 20, 25, 45		35		

How do you find range, median, mode and mean?

The table above shows the scores four students received on the last five math tests in Mr. Wong's class. All four of these students usually have an average of between 95 and 100 in math.

Part A

Analyze the test-score data by completing the table of data.

Part B

Compare the scores the students received on these tests with how they usually score on math tests. What do you think this comparison shows? Justify your answers on the next page.

Test Prep

Name _____

5 Part A

Complete the table of data.

Math Test-Score Data

Tests	Range	Median	Mode	Mean
37, 65, 42, 55, 65	28	55	65	53
58, 42, 75, 85, 45	43	58	—	61
38, 27, 30, 55, 45	28			
45, 35, 20, 25, 45		35		

Part B

Compare the scores the students received on these tests with how they usually score on math tests. What do you think this comparison shows?

Test Taking Tips

How can you check that your answers are accurate?

How can you check that your explanation is clear and complete?

Name _____

1 Katie recorded how much time she spent on homework over a 30-day period. She studied 72 hours during the period. Which equation below could be used to determine the average amount of time she spent on homework each day during the 30-day period?

Ⓐ $t = \dfrac{30}{72}$

Ⓑ $30t = 72$

Ⓒ $72t = 30$

Ⓓ $t = \dfrac{72}{30}$

2 The only vegetable Susan likes is her mother's glazed carrots. Carrots are known to be very good for you. They are especially high in vitamin A. One serving gives Susan four times the amount of the vitamin A she needs each day. However, one serving gives only 12% of the vitamin C she needs.

About how many servings would Susan have to eat to get at least 20% of the vitamin C she needs for the day?

Ⓕ 2

Ⓖ 3

Ⓗ 4

Ⓘ 5

How would you restate the problem in your own words?

Name _____

3 Triangle XYZ is a right triangle. The length of side XY is 3 inches, the length of side YZ is 4 inches.

Find the area of triangle XYZ. Show your work or explain in words how you found the answer.

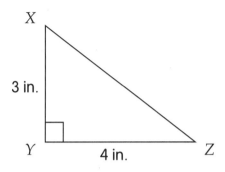

Test Taking Tips

How can you find the area of a triangle?

4

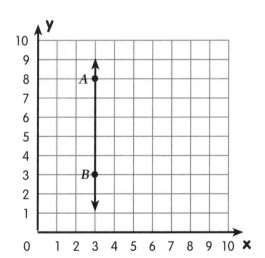

Part A: Identify the coordinates of points A and B on line AB.

Part B: Identify coordinates of two points that lie on a line parallel to line AB. Label the points X and Y. Draw line XY.

Test Taking Tips

What do you need to do to solve the problem? How can knowing the coordinates of line AB help you draw a line parallel to line AB?

Test Prep

89

Name _____

5 Adam is planning to drive from his home in Cleveland, Ohio to Seattle, Washington, a distance of 2,310 miles. He estimates that his car averages 30 miles per gallon of gas and that gas costs about $1.25 per gallon. He plans to drive between 400 miles and 500 miles each day.

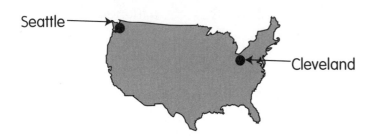

What information does the problem give you?

Part A

About how many gallons of gas will Adam use on his trip? Write an equation and solve.

About how much will the gas for the trip cost? Write an equation and solve.

Part B

How many days will it take Adam to drive

- if he travels 400 miles a day?
- if he travels 500 miles a day?

Write equations to solve.

Part C

If Adam drives 500 miles each day for the first two days and 450 miles each day for the next two days, how many miles will he have left to drive? Draw a diagram to solve.

Name _____

5 Part A

Write an equation that you could use to determine how many gallons of gas Adam will use on his trip. Solve the equation.

_____

_____

Write an equation that you could use to determine how much Adam will spend on gas during his trip. Solve your equation.

_____

_____

Part B

Write equations that you could use to determine how many days it will take Adam to make the trip

- if he drives 400 miles a day.
- if he drives 500 miles a day.

Solve the equations.

_____

_____

Part C

Draw a diagram you could use to help you determine how many miles Adam will have left to drive if he drives 500 miles each day for the first two days and 450 miles each day for the next two days.

Test Taking Tips

How can you check that your answers are accurate?

How can you check that your explanation is clear and complete?

Test Prep

Name _____

1 Jim donates a portion of his earnings to charity every year. This year, for every $100 he earned he donated $3 to charity. His total donation this year was $900.

How much did Jim earn this year?

Ⓐ $9,000

Ⓑ $30,000

Ⓒ $45,000

Ⓓ $90,000

Test Taking Tips

How can writing an equation or writing a proportion help you solve the problem?

---

2 Erin buys 3 raffle tickets in a raffle in which 360 tickets have been purchased. One ticket will be selected for the grand prize.

What is the probability, written as a fraction, that Erin will win the grand prize?

Ⓕ $\frac{1}{130}$

Ⓖ $\frac{1}{360}$

Ⓗ $\frac{1}{120}$

Ⓘ $\frac{357}{360}$

Test Taking Tips

How can you represent the probability of an event occurring using a fraction?

Name _____

Daily Practice Week 23

③ The number of cubic inches in the volume of a cube with edges of 3 in. is 3^3.

Is 3^3 equal to 9? Explain.

_____
_____
_____
_____
_____
_____

Test Taking Tips

How do you compute the area of a square with sides of 3 in.?

How do you compute the volume of a cube with edges of 3 in.?

④ Tina knows the distance around a circle is the circumference. She also knows the formula for finding the circumference is $C = 2\pi r$ where 3.14 is an approximation for π.

Help Tina find the circumference of a circle with a radius of 7 meters. Show your work and explain.

_____
_____
_____
_____
_____
_____

Test Taking Tips

What numbers must Tina multiply together to find the circumference?

Test Prep

Name _____

5 The Acme Storage Company is building a new warehouse. The floor plan below shows the length and width of the building. The building has 14 foot ceilings. The company needs a building with an area of at least 5,000 ft² and a volume of 60,000 ft³.

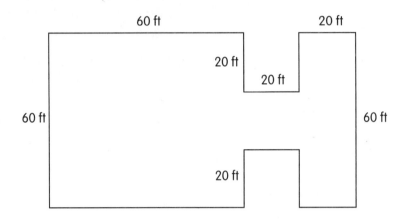

How can knowing the formula for the area of a rectangle help you find the area of the floor of the building?

Part A

What is the area of the building? What is the volume of the building? Does the building meet the area and volume needs of the company? Show your work.

Part B

Draw a different floor plan that would meet the space needs of the Acme Storage Company. Label the dimensions on your plan.

Part C

Explain how you know that your floor plan will meet the space needs of the company.

Test Prep

Name _____

5 Part A

Find the area and the volume of the building shown on the diagram. Be sure to show your work.

area:

_____

volume:

_____

Part B

Draw another floor plan to meet the area needs of the company. Don't forget to label the length and width.

Part C

Explain how you know that your floor plan will meet the space needs of the company.

_____

_____

_____

Test Taking Tips

How can you check that your answers are accurate?

How can you check that your explanation is clear and complete?

Test Prep

Name _____

Daily Practice Week 24

1 Brennan is getting wall-to-wall carpeting in his bedroom and closet. The carpet is sold by the square foot. Below is a floor plan of Brennan's bedroom.

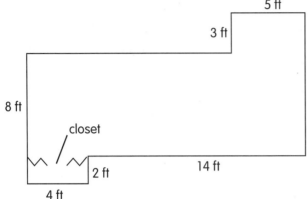

How many square feet of carpet does Brennan need?

Ⓐ 54 ft²

Ⓑ 135 ft²

Ⓒ 150 ft²

Ⓓ 1,200 ft²

Test Taking Tips

How can you use the formula for the area of a rectangle to help you solve the problem?

---

2 Alicia is researching to find what age American presidents were when they took office. She found the following information. George Washington was 57 years old; Thomas Jefferson was 57; Abraham Lincoln was 52; Franklin Roosevelt was 51; John F. Kennedy was 43; and Bill Clinton was 46.

What was the mean age, in years, of these presidents when they took office?

Ⓕ 51

Ⓖ 51.5

Ⓗ 52

Ⓘ 57

Test Taking Tips

What information does the problem give you that can help you solve the problem?

Test Prep

Name _____

3 Karen made a line graph showing her scores on unit tests in English during the year.

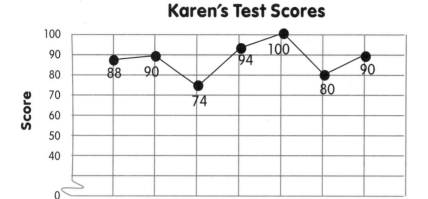

Find the *mean* of Karen's test scores.

_____

In the space below, show your work.

Test Taking Tips

What do you need to solve the problem? Rewrite the problem in your own words.

4 Ali wants to buy a computer program that costs $38. The sales tax rate is 5%. Ali has $40.

Does she have enough money for the program and the sales tax?

_____

Explain.

_____

_____

_____

_____

Test Taking Tips

What does a sales tax rate of 5% mean?

Test Prep

Name _____

5 Jolene recorded these temperature highs and lows for the past four days.

Temperatures in Centerville					
	Mon	Tue	Wed	Thu	Fri
High	23°C	21°C	19°C	18°C	
Low	5°C	8°C	9°C	10°C	

Part A

Use the data in the table to make a multiple-line graph. Be sure to:

- choose an appropriate temperature scale.
- label the axes of the graph.
- write a title.
- include a key.
- mark a point for each high temperature and connect the points.
- mark a point for each low temperature and connect the points.

Part B

If the temperature trends continue, what do you think the temperature high and the temperature low will be on Friday?

Part C

Explain how you decided on your prediction.

Test Taking Tips

What type of graph do you use to show a change over time between two sets of data?

98

Test Prep

Name _____

5

Part A

Make a multiple-line graph of the temperature data in the table. Label your graph and write a title.

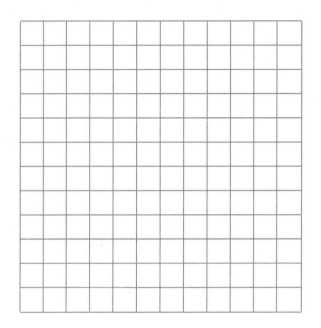

Part B

Predict the temperature high and low on Friday.

_____

Part C

Explain how you decided on your prediction.

_____

_____

_____

_____

_____

Test Prep

Test Taking Tips

How can you check that your answers are accurate?

How can you check that your explanation is clear and complete?

Name _____

1 Charlie ran the mile in 427 seconds.

How many minutes and seconds would that be?

- Ⓐ 5 minutes, 27 seconds
- Ⓑ 6 minutes, 27 seconds
- Ⓒ 7 minutes, 7 seconds
- Ⓓ 8 minutes, 30 seconds

Test Taking Tips

How many seconds are there in one minute?

---

2 Look at point *B* on the graph below. The first number in the ordered pair is 2.

What is the second number in the ordered pair?

- Ⓕ 1
- Ⓖ 2
- Ⓗ 3
- Ⓘ 5

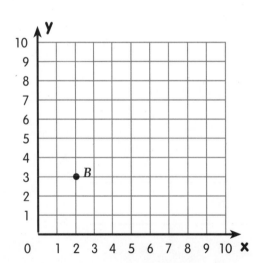

Test Taking Tips

How are the numbers in an ordered pair determined?

100

Test Prep

Name _____

3 The following pattern is made up of regular polygons.

Continue the pattern by drawing the next two figures.

Test Taking Tips

How does the number of sides on each inner figure compare with the number of sides on its outer figure?

---

4 Triangle XYZ is a right triangle. The length of side XY is 3 inches, the length of side XZ is 5 inches.

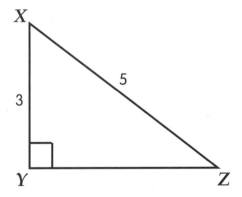

Find the area of triangle XYZ. Show your work and explain in words how you found the answer.

Test Taking Tips

How can knowing the lengths of two sides of a right triangle help you find the length of the third side?

Name _____

5 Jason and his friends like to play football. The line at the center of the field is called the 50-yard line. The lines at the ends of the field are called goal lines.

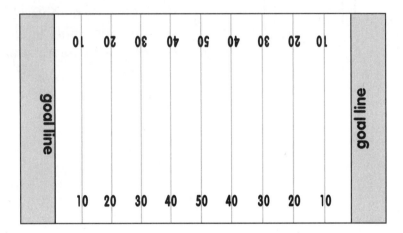

Test Taking Tips

How can you use integers to show the total number of yards gained or lost?

Part A

At one point in the game, Jason's team had the ball at the 50-yard line. They wanted to move toward the goal line at the right. In the next three plays they gained 5 yards, lost 3 yards, and gained 14 yards. Where was the ball after the three plays? Draw a picture or write an explanation to justify your answer.

Part B

Later in the game, Ted's team had the ball on the 50-yard line. They wanted to move toward the goal line at the left. After 3 plays they were 10 yards closer to the goal at the left. During the first play they had lost yards and during the next two plays they gained yards. Describe the number of yards they could have lost and gained in the three plays. Draw a picture and write an explanation to justify your answer.

102

Test Prep

Name _____

Test Taking Tips

5 Part A

Draw a diagram or write an explanation to describe how the ball moved in 3 plays for Jason's team. Don't forget to show where the ball was at the end of the 3 plays.

How can you check that your answers are accurate?

How can you check that your explanation is clear and complete?

Part B

Draw a diagram and write an explanation to describe how the ball could have moved in 3 plays for Ted's team.

Test Prep

Name _____

Daily Practice Week 26

1 Phillip was racing through the grocery store buying some things for his mom. She gave him $10.00. She said he could have any extra money for a candy bar.

As he stood in the checkout line he tossed a candy bar in his basket. As he waited, he estimated the cost of his groceries.

Which is the most reasonable estimate.

4 lb apples	$2.55
5 lb potatoes	$0.98
1 box of cereal	$1.89
1 dozen eggs	$1.29
1 loaf of bread	$1.79
1 can of orange juice	$0.99
1 candy bar	$0.75

Ⓐ Less than $8.00
Ⓑ Between $8.00 and $9.00
Ⓒ Between $9.00 and $10.00
Ⓓ More than $10.00

Test Taking Tips

What do you get when you round each amount to the nearest dollar?

What do you get when you add the rounded amounts?

---

2 Jason is using a metric ruler. He measured his desk as 45 centimeters long. How many millimeters long is his desk?

Ⓕ 4.5 mm
Ⓖ 45 mm
Ⓗ 450 mm
Ⓘ 4,500 mm

Test Taking Tips

How many millimeters are in one centimeter?

104

Test Prep

Name _____

3 Tim is experimenting with ways to purify polluted water. He needs to collect water samples in cylinders that can hold at least 35 in³ of water.

[Cylinder A: 8 in. diameter, 4 in. height]
[Cylinder B: 6 in. diameter, 4 in. height]
[Cylinder C: 4 in. diameter, 2 in. height]
[Cylinder D: 1 in. diameter, 2 in. height]

Which of the cylinders will meet Tim's requirements?

_____

Show your work or explain in words how you found the answer.

_____

_____

Test Taking Tips

What do you need to find out to solve the problem?

---

4 Line AB passes through points (2, 2) and (-2, -2). Line CD passes through points (-2, 2) and (2, -2).

Part A

On a grid, draw lines AB and CD.

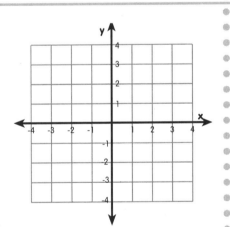

Part B

Name the coordinates of a point where the two lines intersect. _____

Part C

What word could you use to describe the relationship between lines AB and CD?

_____

Test Taking Tips

What angles are formed by the intersection of lines AB and CD?

Test Prep

105

Name _____

Daily Practice Week 26

Test Taking Tips

What will happen if the Marlins win the first 3 games?

5 The playoffs for the Little League division champs are between the Giants and the Marlins. The series is based on the best out of 5. That means that the team that wins the most games out of 5 possible games will be the division champs.

Complete the following on the answer sheet.

Part A

Make an organized list of all possible outcomes. Include wins, losses, and ties.

Part B

What is the least number of games that need to be played to determine the champs? What is the greatest number of games that can be played to determine the champs? Explain your answer.

Part C

Suppose the Marlins won the playoffs in 4 games. One way you can show the games they could have won is to list the winners of each of the 4 games like this:

Marlins, Marlins, Giants, Marlins.

This means the Marlins won games 1, 2, and 4, and the Giants won game 3. List other ways the Marlins could have won the playoffs in 4 games.

List the possible winners in a best out of 5 playoff where the Marlins win in 4 games.

106

Test Prep

Name _____

5 Part A

Make an organized list of all possible outcomes. Include wins, losses, and ties.

Test Taking Tips

How can you check that your answers are accurate?

How can you check that your explanation is clear and complete?

Part B

What is the least number of games played in a best out of 5 playoff?

_____

What is the greatest number of games played?

_____

Explain your answer.

_____
_____
_____

Part C

List the possible winners in a best out of 5 playoff where the Marlins win in 4 games.

_____
_____
_____

Test Prep

107

Name _____

Daily Practice Week 27

1 What is the mean number of children's books sold per day at the Everyday Book Shop for these six days.

Monday	12 books
Tuesday	17 books
Wednesday	12 books
Thursday	12 books
Friday	21 books
Saturday	35 books

Ⓐ about 8 books
Ⓑ about 12 books
Ⓒ about 18 books
Ⓓ about 109 books

Test Taking Tips

How can you eliminate unreasonable answer choices?

2 In math class Charles had to circle all the prime numbers between 1 and 20. He knows there are 8 of them but he only circled 7. Which one does he still need to circle?

1 ② ③ 4 ⑤ 6 ⑦ 8 9 10
⑪ 12 ⑬ 14 15 16 17 18 ⑲ 20

Ⓕ 1
Ⓖ 9
Ⓗ 15
Ⓘ 17

Test Taking Tips

Are there any even numbers that are prime?

108   Test Prep

Name _____

3 Tisha tossed a number cube with sides numbered from 1 to 6. She tossed the cube 18 times. The table shows the outcomes.

Write the mathematical probability of tossing each number.

Compare the mathematical probability to Tisha's experimental probabilities.

Is Tisha's number cube a fair (balanced) cube? Explain your reasoning.

Tisha's Cube Tosses

Possible Outcome	Number of Tosses
1	2
2	3
3	4
4	2
5	4
6	3

Test Taking Tips

Is 18 tosses a reliable sample?

---

4 Mary plotted two points on a coordinate grid. She wrote these ordered pairs for the points.

point A (3, -3)
point B (-3, 0)

Explain the error she made. How would you correct her error?

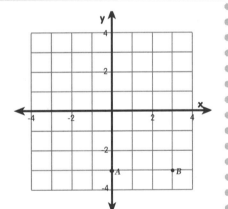

Test Taking Tips

How are the numbers in an ordered pair determined?

_____

_____

_____

_____

Test Prep

109

Name _____

5 Anita deposited $75.00 in a savings account. The account pays 3% simple interest. She plans to leave the money in the bank for 5 years.

Part A

Write an equation to show how much money Anita will have in the account after 1 year. (Remember, she will have the amount she originally deposited plus the interest she has earned on the amount she deposited.)

Solve the equation.

Part B

Complete this table to show the interest that will be earned on Anita's money after it has been in the savings account for 1 year through 5 years.

Anita's Interest					
Years in Account	1	2	3	4	5
Simple Interest Earned	$2.25				

Part C

Describe a pattern you see in the table.

Test Taking Tips

What information does the problem give you?

What do you need to find out to solve the problem?

How do you find simple interest?

Name _____

5 Part A

Write and solve an equation to show how much money Anita will have in her savings account after 1 year.

_____

_____

_____

_____

Part B

Complete the table.

Anita's Interest					
Years in Account	1	2	3	4	5
Simple Interest Earned	$2.25				

Part C

Describe a pattern you see in the table.

_____

_____

_____

_____

_____

Test Taking Tips

How can you check that your answers are accurate?

How can you check that your explanation is clear and complete?

Test Prep

Name _____

Daily Practice Week 28

1 Sandy collects bottles for recycling. The recycling center gives him 3 cents for each large bottle and 2 cents for each small bottle. On his last trip he took 50 large bottles and some small bottles. He received $3.30. Which equation below could be used to determine the number of small bottles that Sandy took to the recycling center?

Ⓐ $0.05y = 3.30$

Ⓑ $0.03(50) + 0.02y = 3.30$

Ⓒ $2y + 3 \times 50 = 3.30$

Ⓓ $3.30(50 + y) = 0.05$

Test Taking Tips

How can you represent "3 cents" as a decimal number?

2 Andy is doing research to find out how many years people keep cars before they replace them. Surveying his neighbors, he gathered the data in this table.

What is the range of the data that Andy collected?

Ⓕ 6

Ⓖ 8

Ⓗ 9

Ⓘ 15

Car Life-Spans	
Car	Years Kept
Taurus	6
Saturn	4
Lexus	6
Civic	4
Bronco	9
Pathfinder	12
Rabbit	3

Test Taking Tips

What is the problem asking you to find?

What information do you need to solve the problem?

Test Prep

Name _____

3 It is predicted that by the year 2000, 30% of the population will be over the age of 60.

Is 30% a rational number? Explain.

_____

_____

_____

_____

_____

_____

_____

Test Taking Tips

What is a rational number?

Can 30% be written as a fraction?

4 Melissa ran 7.5 km. Halfway between the 7.4 km mark and the 7.5 km mark she stopped to retie her shoe. How far had she run before she stopped?

_____

Explain your answer.

_____

_____

_____

_____

Test Taking Tips

How could a number line help you solve this problem?

Test Prep

Name _____

5 Miguel is transforming figures on a coordinate plane to make interesting designs. He starts with triangle ABC with vertices at (2, 1), (5, 1), and (2, 3).

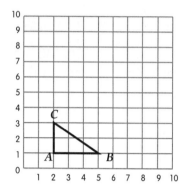

Test Taking Tips

How do you locate a point on a coordinate plane?

Part A

On the grid, draw the coordinate plane shown above. Draw triangle *ABC*. Then reflect triangle *ABC* across the *x*-axis. Draw the new figure. Label its vertices *A'*, *B'*, and *C'*. What are the coordinates of the vertices of the new triangle?

Part B

Compare the coordinates of triangle *ABC* and of triangle *A'B'C'*. Explain how the coordinates are alike and how they are different.

Name _____

5 Part A

On the coordinates below, draw triangle *ABC*. Then reflect triangle *ABC* over the *x*-axis and label the new figure triangle *A'B'C'*.

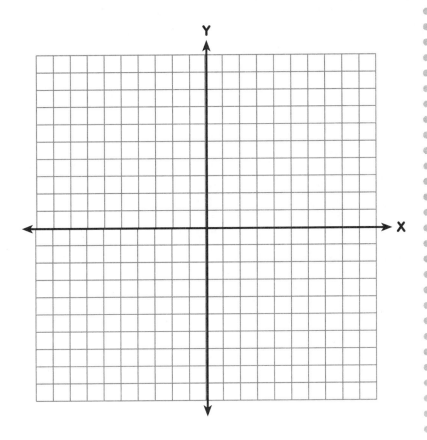

Part B

Compare the coordinates of triangle *ABC* and of triangle *A'B'C'*. Explain how the coordinates are alike and how they are different.

_____

_____

_____

_____

Test Taking Tips

How can you check that your answers are accurate?

How can you check that your explanation is clear and complete?

Name _____

Practice Test 1

1. Rosa has saved twice as much money as her sister and three times as much as her brother. If her brother saved $22.75, how much has Rosa saved?

   Ⓐ $44.50

   Ⓑ $45.50

   Ⓒ $66.50

   Ⓓ $68.25

2. The stock market had a big day. It rose 304.75 points. What is the correct written form for this number?

   Ⓕ Three hundred forty and seventy-five hundredths

   Ⓖ Three hundred four and seventy-five

   Ⓗ Three hundred four and seventy-five thousandths

   Ⓘ Three hundred four and seventy-five hundredths

Test Prep

Name _____

Practice Test 1

3 Jason said that 33 students were going to the picnic in the park. If there are 100 students in the class, what percent of students in the class would be going to the park?

- Ⓐ 33%
- Ⓑ 67%
- Ⓒ 70%
- Ⓓ 100%

4 Anthony and his dad were out to test drive a new gas-saving car his dad bought. They drove 430 miles to Atlanta, Georgia. If the car used 20 gallons of gas for the trip, how many miles per gallon were they getting?

- Ⓕ 21.5
- Ⓖ 215
- Ⓗ 450
- Ⓘ 8,600

Name _____

Practice Test 1

5 The graph shows the number of girls and boys in Mr. Ito's sixth-grade class. There are 5 students in Mr. Ito's class who wear glasses. What percent of the students wear glasses?

Ⓐ 20%

Ⓑ 25%

Ⓒ 33%

Ⓓ 50%

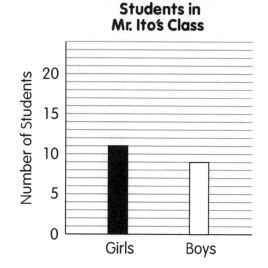

6 Juanita saved $100.00 to go on a shopping spree. She spent $42.27 on a new dress, $27.55 on a new pair of pants, $10.53 on a new CD. If she left the house with $100.00, how much of her money is left?

Ⓕ $19.55

Ⓖ $19.65

Ⓗ $19.75

Ⓘ $20.75

Test Prep

Name _____

Practice Test 1

1. $3 \times 3 \times 3 = 3^3 = 27$
2. $2 \times 2 \times 2 = 2^3 = 6$
3. $4 \times 4 \times 4 \times 4 = 4^4 = 256$
4. $5 \times 5 \times 5 \times 5 = 5^4 = 625$

The first 4 problems of Allen's math homework are written above. He is learning to write numbers in exponent form. Three of the problems are correct. One problem is incorrect. Find the incorrect problem and write it correctly. What was Allen doing wrong?

_____
_____
_____
_____

Alex knows the multiples of a number could go on forever. For example, the multiples of 4 are:
4, 8, 12, 16, 20, 24, 28, 32, 36,___, ___, ___
Tell what the next 3 multiples of 4 are. Explain how you found them.

_____
_____
_____
_____
_____
_____
_____

Test Prep

119

Name _____

Practice Test 1

Mrs. Jones is planting a garden in her yard. She is making it 15 feet wide and 30 feet long. Mrs. Jones needs to know the perimeter of the garden in order to build a fence to keep out the rabbits. She also needs to know the area of the garden so she can buy the correct amount of topsoil.

Part A

Draw a diagram of Mrs. Jones' garden. Find the perimeter and area of the garden you have drawn.

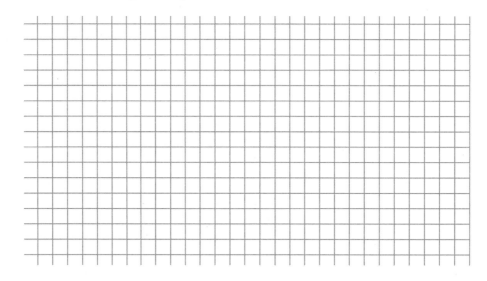

Name _____

Practice Test 1

Part B

Draw a diagram of another rectangular garden that has the same perimeter as Mrs. Jones' garden but a different area.

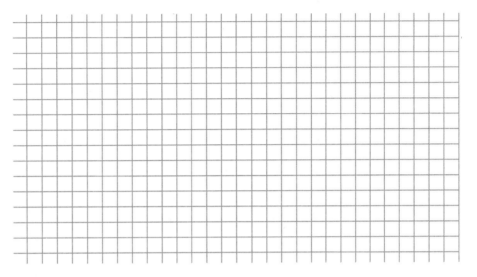

Part C

Could you use the same amount of fencing to enclose your garden as Mrs. Jones did to enclose hers? Justify your answer.

_____

_____

_____

_____

_____

_____

Name _____

Practice Test 1

10 Mr. Carpenter returned the last math test to the students in his class. Scores went from 85 all the way down to 32. He explained that the range was very wide. What is the range of scores for this math test?

    Ⓐ 32
    Ⓑ 52
    Ⓒ 53
    Ⓓ 85

11 David needs 3 gallons of lemonade for his party. The pitcher he is using to make the lemonade holds 2 quarts. How many pitchers of lemonade does he need to make?

    Ⓕ 6 pitchers
    Ⓖ 8 pitchers
    Ⓗ 10 pitchers
    Ⓘ 12 pitchers

Name _____

Practice Test 1

12. The information in the graph below shows the progress of 4 companies in sales for 1997. According to the graph, which company had the most sales in the 4th quarter?

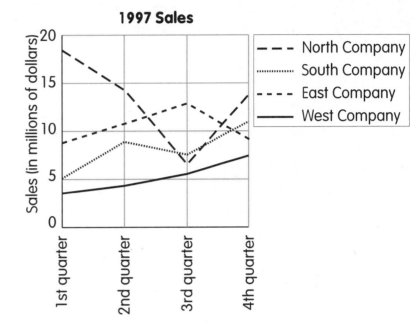

Ⓐ North
Ⓑ South
Ⓒ East
Ⓓ West

13. Allen is making a card that measures 6 cm long by 4 cm wide. He is covering the entire card with 1 cm by 1 cm animal stamps that he has collected. How many stamps can he put on the card so that none overlap and so that there are no gaps?

Ⓕ 18 stamps
Ⓖ 20 stamps
Ⓗ 24 stamps
Ⓘ 30 stamps

Allen's Card

Test Prep

123

Name _____

Practice Test 1

14. Trevor has a small basketball court in his yard. It is 20 feet wide and 20 feet long. It has an area of 400 square feet. His dad said they could pour more concrete and double the length of the court but the width would remain the same. If they double the length and leave the width the same, what will the area of the new court be?

   A 440 ft²

   B 600 ft²

   C 800 ft²

   D 1,600 ft²

15. Anthony needs to roll a 4 on the number cube in order to win the game he is playing. The number cube is labeled 1, 2, 3, 4, 5, and 6. He is discouraged because he has rolled a 4 on his last 3 turns and thinks he can't do it again when he needs to. What is the probability, written as a fraction, that he will roll a 4 on his next turn.?

   F $\frac{1}{6}$

   G $\frac{1}{4}$

   H $\frac{1}{3}$

   I $\frac{1}{2}$

Name _____

Practice Test 1

Notes

Problems that I answered correctly.

_____

_____

_____

_____

_____

_____

_____

Problems that I did not understand.

_____

_____

_____

_____

_____

Vocabulary that I need to learn.

_____

_____

_____

_____

_____

Name _____

Practice Test 2

1. One ounce is six hundred twenty-five ten-thousandths of a pound. Which is the correct written form of this number?

   A   6250.0000

   B   625.0000

   C   0.6250

   D   0.0625

2. The temperature in New York City dropped to ⁻4°F. Which kind of number expresses this temperature?

   F   integers

   G   whole numbers

   H   fractions

   I   decimals

Name _____

Practice Test 2

 3 If 80% of the students bought hot lunch on Thursday, what fraction of the students did NOT buy hot lunch?

- Ⓐ $\frac{1}{5}$
- Ⓑ $\frac{1}{4}$
- Ⓒ $\frac{4}{5}$
- Ⓓ $\frac{8}{10}$

 4 Thomas has $\frac{3}{4}$ of the money he saved in a savings account at the bank. What percent of his money is in the savings account?

- Ⓕ 20%
- Ⓖ 25%
- Ⓗ 50%
- Ⓘ 75%

Name _____

Practice Test 2

5. Christopher's dad is a scientist. He likes to speak in scientific terms. He told Christopher that Hawaii is about 10^3 miles away. About how many miles away is Hawaii?

Ⓐ 30 miles

Ⓑ 100 miles

Ⓒ 1,000 miles

Ⓓ 10,000 miles

6. There are 150 students in the auditorium watching the play practice. Joel said $\frac{2}{3}$ of them missed the last bus and had to find another way home. How many students missed the bus?

Ⓕ 50 students

Ⓖ 75 students

Ⓗ 80 students

Ⓘ 100 students

Test Prep

Name _____

Practice Test 2

7 Kristin and Lisa's math teacher, Mr. Jones, has a jar filled with candy for Halloween. Mr. Jones promised that on Halloween he will put all his students' names in a hat and draw one to be the winner of the candy jar. If there are 28 students in the class, what is the probability that either Kristin or Lisa will win? Explain your answer.

_____
_____
_____
_____
_____

8 There is a nonstop flight from Tampa, Florida, to Cincinnati, Ohio. The flight leaves at 7:50 A.M. and arrives in Cincinnati at 9:50 A.M. (Both times are Eastern Standard Time.) If the air distance is 772 miles, how fast is the jet flying? Show your work and explain your answer.

_____
_____
_____
_____
_____

Test Prep

Name _____

Practice Test 2

Lisa kept track of the daily high temperatures for one week in Hawaii because she wanted to convince her dad that they needed to go there for a winter vacation. Starting February 7, the daily high temperatures in a city in Hawaii were 77°F, 86°F, 75°F, 78°F, 80°F, 82°F, 85°F. Lisa's dad said if she would graph this information it might be more convincing.

Part A

What type of graph can Lisa use to graph these changes over 7 days?

Part B

Explain why the type of graph you chose was appropriate.

Practice Test 2

Name _____

9 Part A

Make a graph that best shows the changes in temperature. Be sure to label your graph and write a title.

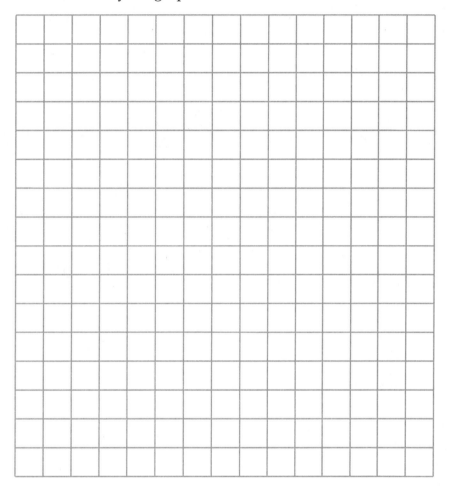

Part B

Explain why the type of graph you chose is the most appropriate to show the data.

_____

_____

_____

_____

Test Prep

Name _____

Practice Test 2

10 George needed to figure out the coordinates for point B on the grid below. Look at point B. Tell what the correct coordinates are for identifying that point.

Ⓐ (⁻4, 4)

Ⓑ (4, ⁻4)

Ⓒ (0, ⁻4)

Ⓓ (⁻4, 0)

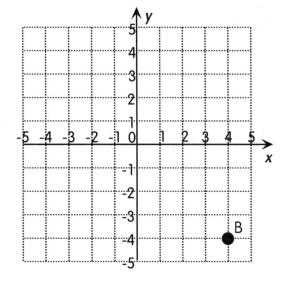

11 Rick was 12 and his little brother Tad was 6. Since Rick was twice his brother's age, he thought he should have twice as much allowance. If his brother got $2.00, then he should get $4.00, and if his brother got $4.00, then he should get $8.00. If *t* represents Tad's allowance, then which expression below best shows Rick's idea?

Ⓕ t + 2

Ⓖ 2 x t

Ⓗ t + $ 4.00

Ⓘ t x 4

132

Test Prep

Name _____

Practice Test 2

 Find the pattern in the table below. For each row, find how the value for y is related to the value for x. What is the value of r?

x	y
4	1
7	4
10	7
13	10
16	r

Ⓐ 11
Ⓑ 13
Ⓒ 14
Ⓓ 15

 Jason wants to figure out how much a star professional athlete makes in a month. Rounded to the nearest dollar, how much does an athlete earn per month if he or she earns a million dollars a year?

Ⓕ $83,330
Ⓖ $83,333
Ⓗ $83,334
Ⓘ $83,340

Test Prep

Name _____

Practice Test 2

There are 4 children waiting in line to ride the elephant at the zoo. The zoo keeper said that on weekends he has seen that line double every 5 minutes during the first hour the zoo is open. If that were true, how many children would be waiting in line after 20 minutes?

Number of minutes	5	10	15	20
Number of children	4	8	16	?

- Ⓐ 20 children
- Ⓑ 22 children
- Ⓒ 24 children
- Ⓓ 32 children

West Middle School is building a new gymnasium. Students can submit scale drawings to share their ideas for a floor plan. Tanya started to make a scale drawing. Her scale is 1 cm = 10 ft and she wants the new gymnasium to be 60 feet long. How many centimeters should she make the length of her scale drawing of the gymnasium?

- Ⓕ 5 cm
- Ⓖ 6 cm
- Ⓗ 60 cm
- Ⓘ 600 cm

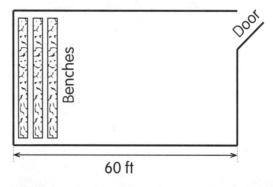

Gymnasium/Auditorium

134

Test Prep

Name _____

Practice Test 2

Notes

Problems that I answered correctly.

_____
_____
_____
_____
_____
_____

Problems that I did not understand.

_____
_____
_____
_____

Vocabulary that I need to learn.

_____
_____
_____
_____

Name _____

Practice Test 3

Statistics show that nine out of ten people are right-handed. If this is correct, what percent of people are left-handed?

- Ⓐ 10%
- Ⓑ 70%
- Ⓒ 80%
- Ⓓ 90%

In exponent form 1 million (1,000,000) is written 10^6. Which shows 10 million (10,000,000) in exponent form?

- Ⓕ 10^5
- Ⓖ 10^6
- Ⓗ 10^7
- Ⓘ 10^8

Name _____

Practice Test 3

 Which figures below show an example of a reflection of the triangle?

Ⓐ

Ⓑ

Ⓒ

Ⓓ

 There are 50 students in the 6th grade. If $\frac{1}{2}$ of the students are girls, what percent are girls?

- Ⓕ 10%
- Ⓖ 20%
- Ⓗ 25%
- Ⓘ 50%

Test Prep

Name _____

Practice Test 3

  About $\frac{1}{4}$ of the students at West Middle School bring lunch from home. This is represented in the graph below. What percent of students bring their lunch from home?

Ⓐ 20%

Ⓑ 25%

Ⓒ 50%

Ⓓ 75%

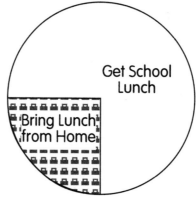

  Chris didn't understand his math problem.

3 + 5 × 10 = ?

He kept getting the same answer of 150 and he kept getting it marked wrong. What is the correct answer?

Ⓕ 18

Ⓖ 53

Ⓗ 80

Ⓘ 450

Name _____

Practice Test 3

7. Aaron is helping his mom prepare snacks for her preschool class. The 10 children are having oranges for a snack. Each child gets 6 sections of an orange. Aaron peeled one orange and found 14 sections.

If all the oranges have the same number of sections as the first one he peeled, about how many oranges will Aaron have to peel to have enough sections for the children?

Show your work. Explain your reasoning.

_____

_____

_____

_____

8. Drew left food outside after the picnic. Lots of ants were gathering around the food. The exterminator said he estimated that there were about 100 ants in the colony now. He suggested that if the colony was left alone, the ant population would double every day.

Complete the table to show how many ants there would be in the colony by day 4 and day 5.

Explain your reasoning.

Ant Colony Population

Day	Number of Ants
day 1 (today)	100
day 2	200
day 3	400
day 4	
day 5	

Test Prep

139

Name _____

Practice Test 3

Zack is helping his dad put some new sod in their back yard. His dad measured the area for the new sod. It measured 44 feet long and 60 feet wide. His dad can buy sod for $0.75 a square foot.

Part A

How many square feet of sod will Zack's dad need?

Part B

How much will the sod cost?

Show your work and justify your answers.

Name _____

Practice Test 3

Part A

How many square feet of sod will Zack's dad need?

Part B

How much will the sod cost?

Show your work and justify your answers.

_____
_____
_____
_____
_____
_____

Test Prep

141

Name _____

Practice Test 3

At Pete's Pumpkin Patch pumpkins are priced according to their diameters. Pete says the diameter of a pumpkin is proportional to its weight.

If Pete is correct and if a pumpkin with a 20 centimeter diameter weighs 3.5 pounds, how much will a pumpkin with a diameter of 60 centimeters weigh?

- Ⓐ 3.5 pounds
- Ⓑ 5 pounds
- Ⓒ 6 pounds
- Ⓓ 10.5 pounds

Susan is measuring fabric to make a costume for the class play. She needs 4 yards and she is using a foot ruler because she can't find a yardstick.

If she has already measured 4 feet of the fabric, how many more feet does she need to measure to get 4 yards?

- Ⓕ 6 feet
- Ⓖ 8 feet
- Ⓗ 9 feet
- Ⓘ 12 feet

Test Prep

Name _____

Practice Test 3

12 Jared is building a square box for his little rabbit to live in. He knows a square has 4 angles. What is the measure of each angle in a square?

- Ⓐ 45°
- Ⓑ 90°
- Ⓒ 180°
- Ⓓ 360°

13 Chris's coach said he could play in the next basketball game if the average of his math scores is at least 75. His math scores for the past week were 55, 82, 75, 92, 100.

What is his mean (average) math score for the week?

- Ⓕ 75
- Ⓖ 78.8
- Ⓗ 80.8
- Ⓘ 82

Test Prep

143

Name _____

Practice Test 3

 Wilson was playing "Roll it Out" in math class. In order for him to win the game he needed to roll a 3 on the number cube that was labeled 1, 2, 3, 4, 5, and 6.

What is the probability, written as a fraction, that he will roll a 3?

- Ⓐ $\frac{1}{6}$
- Ⓑ $\frac{1}{3}$
- Ⓒ $\frac{1}{2}$
- Ⓓ $\frac{2}{3}$

 Find the pattern in the table.

What is the value of *t*?

- Ⓕ 17
- Ⓖ 20
- Ⓗ 25
- Ⓘ 30

x	y
1	5
2	10
3	15
4	
5	t

144

Test Prep

Name _____

Answer Sheet • page 1

Practice Test ____

1. Ⓐ Ⓑ Ⓒ Ⓓ
2. Ⓕ Ⓖ Ⓗ Ⓘ
3. Ⓐ Ⓑ Ⓒ Ⓓ
4. Ⓕ Ⓖ Ⓗ Ⓘ
5. Ⓐ Ⓑ Ⓒ Ⓓ
6. Ⓕ Ⓖ Ⓗ Ⓘ

Name _____

Answer Sheet • page 2

7 _____

_____

_____

_____

_____

_____

_____

_____

_____

_____

8 _____

_____

_____

_____

_____

_____

_____

_____

_____

_____

Name _____

Answer Sheet • page 3

9

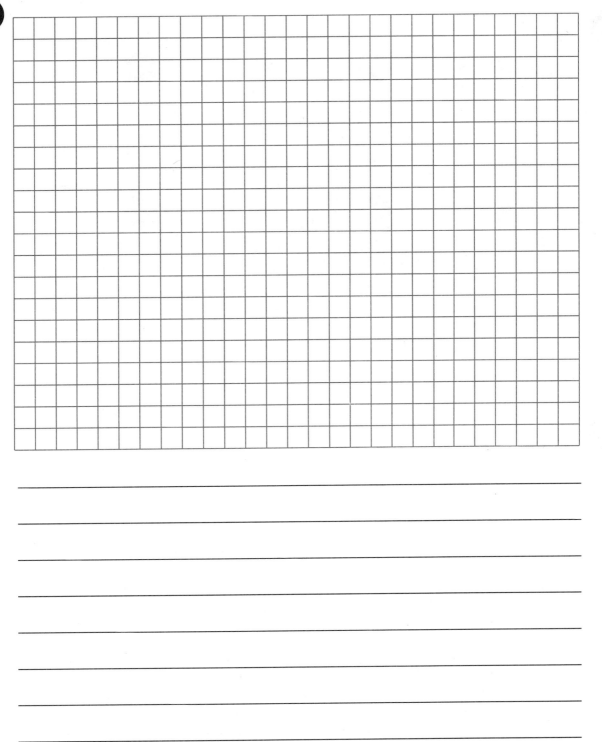

Name _____

Answer Sheet • page 4

Practice Test ____

10. Ⓐ Ⓑ Ⓒ Ⓓ
11. Ⓕ Ⓖ Ⓗ Ⓘ
12. Ⓐ Ⓑ Ⓒ Ⓓ

13. Ⓕ Ⓖ Ⓗ Ⓘ
14. Ⓐ Ⓑ Ⓒ Ⓓ
15. Ⓕ Ⓖ Ⓗ Ⓘ

STOP

Name _____

Answer Sheet • page 1

Practice Test ____

1 Ⓐ Ⓑ Ⓒ Ⓓ   2 Ⓕ Ⓖ Ⓗ Ⓘ   3 Ⓐ Ⓑ Ⓒ Ⓓ

4 Ⓕ Ⓖ Ⓗ Ⓘ   5 Ⓐ Ⓑ Ⓒ Ⓓ   6 Ⓕ Ⓖ Ⓗ Ⓘ

Test Prep • Answer Sheet

Name _____

Answer Sheet • page 2

Daily Practice

Practice Test ____

7 _____

8 _____

Name _____

Answer Sheet • page 3

Practice Test ____

9

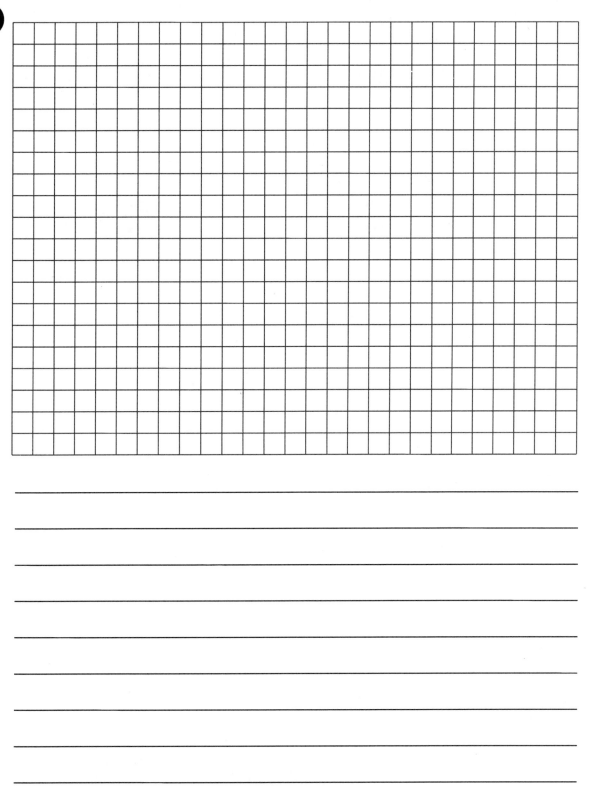

_____

_____

_____

_____

_____

_____

_____

Name _____

Answer Sheet • page 4

Practice Test ____

10 Ⓐ Ⓑ Ⓒ Ⓓ   11 Ⓕ Ⓖ Ⓗ Ⓘ   12 Ⓐ Ⓑ Ⓒ Ⓓ

13 Ⓕ Ⓖ Ⓗ Ⓘ   14 Ⓐ Ⓑ Ⓒ Ⓓ   15 Ⓕ Ⓖ Ⓗ Ⓘ

STOP

Name _____

Answer Sheet • page 1

Practice Test ____

1. Ⓐ Ⓑ Ⓒ Ⓓ   2. Ⓕ Ⓖ Ⓗ Ⓘ   3. Ⓐ Ⓑ Ⓒ Ⓓ

4. Ⓕ Ⓖ Ⓗ Ⓘ   5. Ⓐ Ⓑ Ⓒ Ⓓ   6. Ⓕ Ⓖ Ⓗ Ⓘ

Name _____

Answer Sheet • page 2

Practice Test ____

7. _____

8. _____

Name _____

Answer Sheet • page 3

Practice Test ____

9

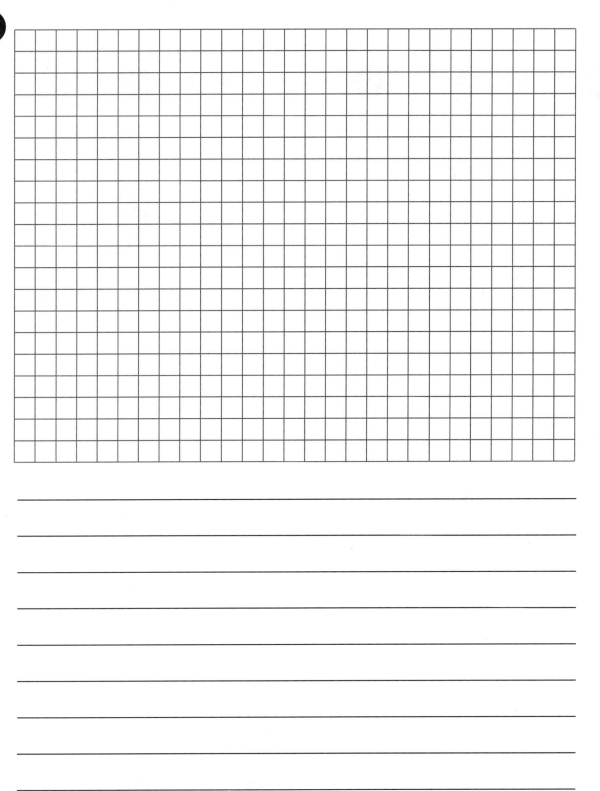

_____
_____
_____
_____
_____
_____
_____
_____

Name _____

Answer Sheet • page 4

Practice Test ____

10 Ⓐ Ⓑ Ⓒ Ⓓ   11 Ⓕ Ⓖ Ⓗ Ⓘ   12 Ⓐ Ⓑ Ⓒ Ⓓ

13 Ⓕ Ⓖ Ⓗ Ⓘ   14 Ⓐ Ⓑ Ⓒ Ⓓ   15 Ⓕ Ⓖ Ⓗ Ⓘ

STOP

Name _____

Mathematics Reference Sheet

Formulas

 Triangle   Area = $\frac{1}{2}bh$

 Rectangle   Area = lw

 Trapezoid   Area = $\frac{1}{2}h(b_1 + b_2)$

 Parallelogram   Area = bh

 Circle   Area = πr^2
Circumference = $\pi d = 2\pi r$

Key
b = base
h = height
l = length
w = width

d = diameter
r = radius

Use 3.14 or $\frac{22}{7}$ for π.

In a polygon, the sum of the measures of the interior angles is equal to $180(n-2)$, where n represents the number of sides.

Pythagorean Property:
$c^2 = a^2 + b^2$

 Right Circular Cylinder   Volume = $\pi r^2 h$ Total Surface Area = $2\pi rh + 2\pi r^2$

 Rectangular Solid   Volume = lwh Total Surface Area = $2(lw) + 2(hw) + 2(lh)$

Conversions

1 yard = 3 feet = 36 inches
1 mile = 1,760 yards = 5,280 feet
1 acre = 43,560 square feet
1 hour = 60 minutes
1 minute = 60 seconds

1 liter = 1,000 milliliters = 1,000 cubic centimeters
1 meter = 100 centimeters = 1,000 millimeters
1 kilometer = 1,000 meters
1 gram = 1,000 milligrams
1 kilogram = 1,000 grams

1 cup = 8 fluid ounces
1 pint = 2 cups
1 quart = 2 pints
1 gallon = 4 quarts

1 pound = 16 ounces
1 ton = 2,000 pounds

Test Prep • Reference Page